Groundwater Modeling Using Geographical Information Systems

Groundwater Modeling Using Geographical Information Systems

George F. Pinder
University of Vermont

John Wiley & Sons, Inc.

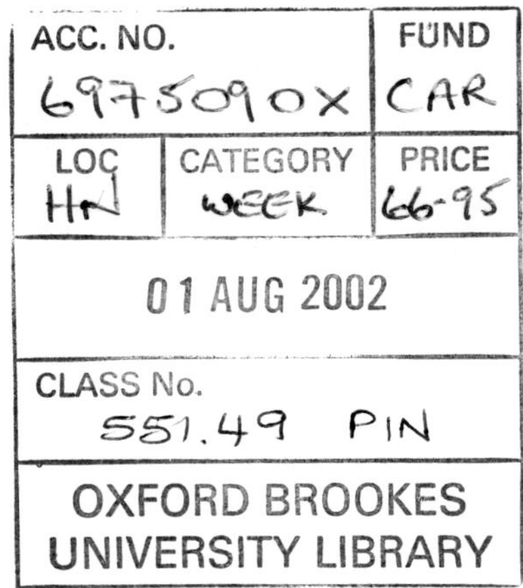

This book is printed on acid-free paper. ♾

Library of Congress Cataloging-in-Publication Data

Pinder, George Francis, 1942–

Groundwater modeling using geographical information systems / George F. Pinder.

p. cm.

ISBN 0-471-08498-0 (alk. paper)

1. Groundwater flow—Mathematical models. 2. Geograhic information systems. I. Title.

GB1197.7.P55 2003

551.49′01′1–dc21 2002004929

Printed in the United States of America

10 9 8 7 6 5 4 3 2 1

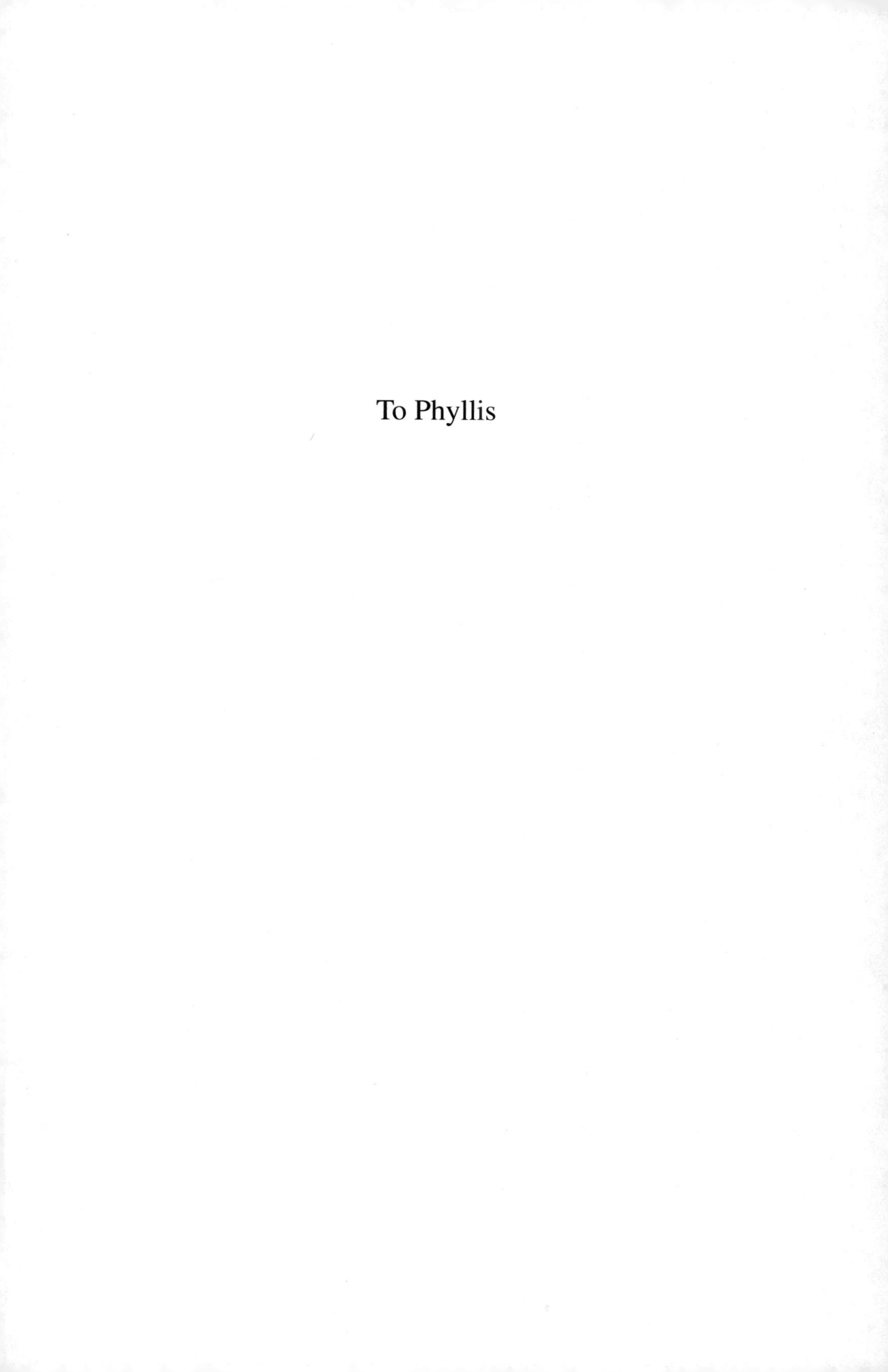

To Phyllis

Contents

Preface

The purpose of this book is to present elements of the art of groundwater flow and transport modeling using tools generally identified with geographical information systems (GISs). The book is the outgrowth of notes I prepared for teaching a course in groundwater flow and transport modeling while I was teaching at Princeton University. The concept of employing GIS as an integral part of the modeling course was added during my tenure at the University of Vermont.

The motivation for introducing a GIS format in the course stems from the realization that from the outset, groundwater modeling has entailed the organization, quantification, and interpretation of large quantities of geohydrological data. Early work in groundwater modeling required the translation and transfer of information on maps, charts, and tables into computer-readable form. The work was lengthy, tedious, and error prone. Changes that were required in the data sets in the course of calibrating the models often involved sifting through thousands of numbers to make what often turned out to be minor modifications to the input-data sets.

The specification of hydrological information such as rainfall, parametric information such as hydraulic conductivity, design parameter specifications such as well locations and discharge values, and auxiliary conditions such as boundary conditions all involve the organization and manipulation of enormous quantities of data. Virtually all of this information is spatially, and in some instances temporally, distributed. Much of it is available in computerized databases either as maps in bitmap or vector image format or as data tables. Due to advances in computer-graphical technology, the information in such databases is now accessed most efficiently through GIS systems.

The resulting groundwater model-building tools generally incorporate a Windows-based, user-friendly, graphically oriented, functionally integrated, data-input, analysis, and postprocessing system. I have used two such systems in this book to facilitate presentation of the basic concepts of groundwater flow and transport modeling. In each instance the system consists of the Argus ONE Geographic Information Modeling (GIM) environment, a groundwater flow and transport model, and a plug-in extension (PIE) that interfaces Argus ONE and the model.

The Princeton Transport Code (*PTC*), *MODFLOW*, and *MT3D* are the groundwater flow and transport models discussed in this book. These three models were selected from a universe of possible candidates because (1) they are widely used in practical application; (2) collectively, they represent both the finite-difference and finite-element numerical-modeling approaches; and (3) plug-in extensions (PIEs) have been developed and are available at http://www.argusint.com.

Using the GIS approach, the analyst works with the original spatial information: for example, information provided on maps. Such information is generally accessible and is normally cataloged and presented in commonly understood terminology rather than in the more specialized vocabulary of the groundwater-modeling professional. A visually based, computer-graphical approach, this method of data organization and analysis is much more intuitive than cumbersome utilization of numerical arrays. I refer to the above-described GIS approach as the *geographic modeling approach* (GMA).

The book consists of three parts. Part 1 is dedicated to groundwater-flow modeling, Part 2 to groundwater-transport modeling, and Part 3 is a model-development tutorial that considers both finite-difference- and finite-element-based approaches. A comparison of these two approaches is also provided in this part.

The *PTC* used extensively in the preparation of this manuscript was developed over a period of approximately 20 years. Among those besides myself who have contributed to its development are D. P. Ahlfeld, D. K. Babu, L. R. Bentley, E. O. Frind, J. F. Guarnaccia, G. P. Karatzas, A. Niemi, R. H. Page, M. P. Papadopoulou, A. A. Spiliotopoulos, S. A. Stothoff, and K. Yamada. The *MODFLOW* and *MT3D* computer codes employed in this text are provided by the U.S. Geological Survey. The Argus ONE site (http://www.argusint.com) provides links to these models and their associated PIEs. The PIE for *PTC* was created by J. L. Olivares. *PTC*, its documentation, and the PIE interface to Argus ONE can be downloaded from www.wiley.com/go/pinder.

I am indebted to those who provided helpful criticisms and contributions during the preparation of this manuscript. Of special note are F. Fedele, M. C. McKay, J. Margolin, T. Mascarenhas, M. M. Ozbek, M. P, Papadapoulou, K. L. Ricciardi, X. Wei, and Y. Zhang.

George F. Pinder
Burlington, Vermont

Groundwater Modeling Using Geographical Information Systems

Part 1

Flow Modeling

1.1 INTRODUCTION

Over the past three decades, groundwater-flow and transport modeling has evolved from a scientific curiosity to a widely used design and analysis technology. At the outset, groundwater modeling focused on the evaluation of groundwater supplies from the perspective of quantity, but more recent applications have addressed issues of water quality. Groundwater resource issues involving primarily water quantity are largely addressed by groundwater-flow models. Groundwater-transport models, however, are often needed when the problem to be addressed involves groundwater quality. A groundwater-flow model is a necessary precursor to the development of a groundwater-transport model. The groundwater velocity needed in the transport model is obtained from the flow model.

Groundwater flow models have a long history and come in many forms. Early flow models were based primarily on the finite-difference method of approximation of the governing field equations. Simple in concept and computationally efficient, finite-difference models found broad acceptance by the groundwater community. Later model development focused on the finite-element approach, which was more mathematically abstract and more difficult to code. The finite-element approach had the advantage of being able to represent irregular aquifer geometries more accurately because unlike the broadly used version of the finite-difference model which relied on rectangular meshes,[1] finite-element models could accommodate triangular and even deformed rectangular meshes. Both finite-difference and finite-element models

[1] Early finite-difference models were also available that could accommodate polygonal meshes, but they were not widely used.

are currently used routinely in groundwater hydrology and groundwater-contaminant hydrology to predict groundwater-reservoir behavior.

In this chapter we provide, through a field example, the conceptualization, formulation, and construction of a groundwater-flow model. Model construction, whether based on finite-difference or finite-element methods of approximation, involves a number of well-defined steps. In summary, these steps are as follows:

1. Establish the minimum area to be represented by the model;
2. Determine the hydrological features that can serve as boundaries to the model.
3. Compile the geological information.
4. Compile the hydrological information.
5. Determine the number of physical dimensions needed for the model.
6. Define the size of the model.
7. Define the model discretization.
8. Input the model boundary conditions.
9. Input the model parameters.
10. Input the model stresses.
11. Run the model.
12. Output the calculated hydraulic heads.
13. Calibrate the model.
14. Make the production runs.

To clarify the various aspects of modeling, we will introduce a field site located in Tucson, Arizona. Using the Argus ONE modeling environment,[2] we will illustrate each of the steps listed above. In this example we focus on the contaminant trichloroethylene (**TCE**), the major contaminant of concern (**COC**) at this site.

Tucson Example

As an introduction to the Tucson site, we provide the following description recorded by one of the groundwater professionals who investigated the area (Rampe [4])

> Groundwater pollution in the vicinity of the Tucson, Arizona, International Airport has been known or suspected since the early 1950's. At that time, although some drinking water wells had been affected, the full extent of the pollution was not investigated. In some measure this appears to have been due to efforts on the part of government and industry to control the effects of groundwater pollution by controlling the above-ground pollution sources and by providing alternate supplies to those affected. It is also possible that the implications for the presence of very low levels of organic pollutants in drinking water were not fully appreciated by those involved at the time. In 1981, extensive

[2]Argus ONE is a commercially available program. It is a programmable interface that allows one to access the PTC groundwater code as well as other groundwater modeling codes in a Windows environment.

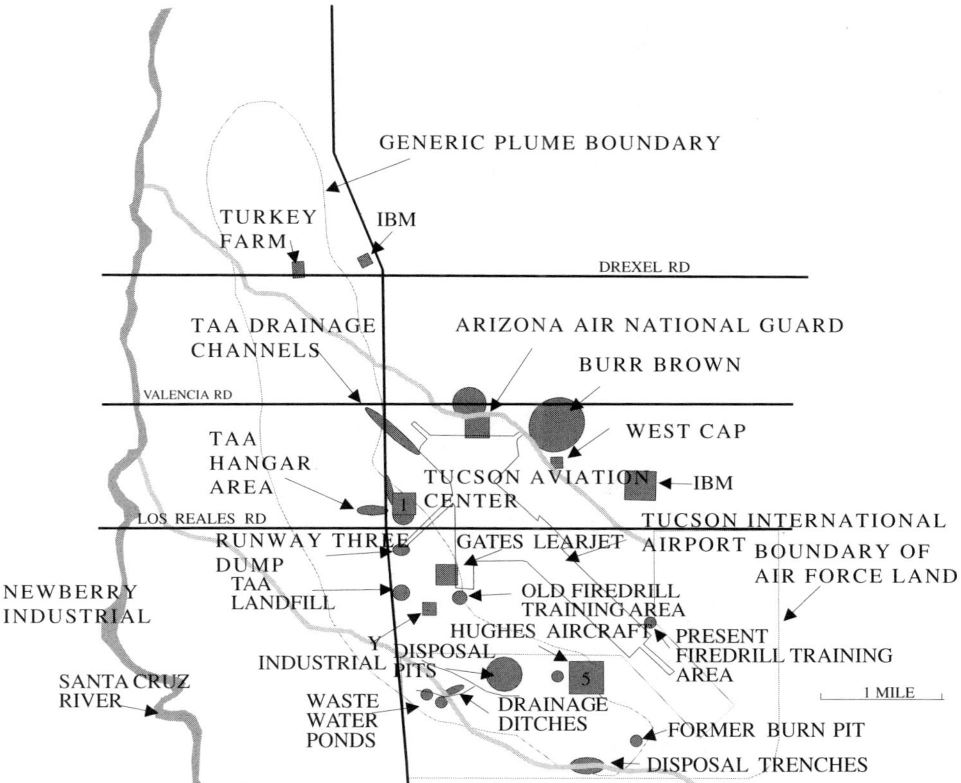

FIGURE 1.1. *A potential contaminant source from a regional perspective (modified from Black and Veatch [10].*

groundwater contamination by *volatile organic compounds* (**VOC**s) was discovered. The most abundant pollutant found was *trichloroethylene* (**TCE**), which has since been shown to occur in an area roughly extending from the Hughes Aircraft Company facility (**HAC**) in a northwesterly direction to Irvington Road (Figures 1.1 and 1.2). Other contaminants have also been found in the plume. These include chromium, isomers of *dichloroethylene* (**DCE**), *benzene*, *chloroform*, and other organic compounds. TCE is a compound suspected of being a carcinogen by the National Research Council, and has been placed on the Environmental Protection Agency's (**EPA**) list of *Priority Pollutants* [5]. The *Arizona Department of Health Services* (**ADHS**) adopted an action limit of 5 **ppb**[3] (parts per billion) for TCE in drinking water supplies.

The extent and severity of the contamination prompted action from Federal agencies. The U.S. Air Force began investigations of groundwater conditions in 1981, and in 1982 embarked upon a program of aquifer restoration south of Los Reales Road. *The*

[3] *ppb* is the acronym for parts per billion, which is the weight in grams of a compound per billion grams of solution.

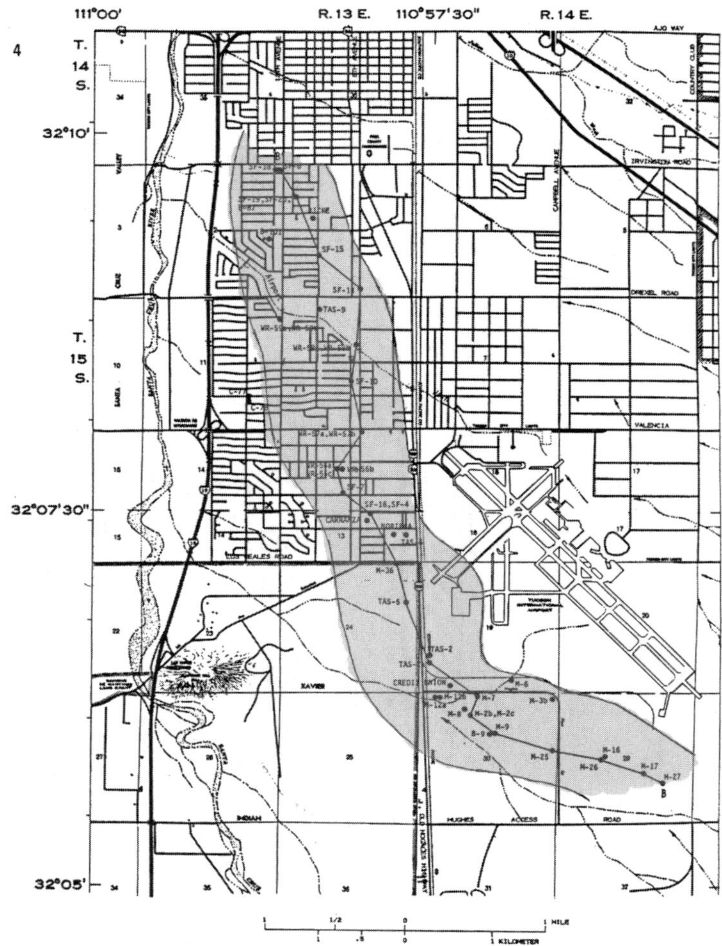

FIGURE 1.2. *Approximate area of trichloroethylene (TCE) contamination in groundwater near the Tucson airport (modified from Leake and Hanson [7]).*

> *Tucson Airport Area* (**TAA**) was placed in Superfund's original *National Priority List* in 1982. EPA began investigations under Superfund to investigate further the sources and occurrence of groundwater contamination north of Los Reales Road. ADHS applied for and received funding from EPA under a Superfund cooperative agreement, and was named lead agency for the TAA.

After an extensive discussion of the scope, goals, and objectives of the investigation as well as the methodology employed, Rampe [4] provides the following conclusions:

> Several potential sources of groundwater contamination exist in the vicinity of the Tucson International Airport. These are summarized in Table 1.1. Based on evidence gath-

TABLE 1.1. Summary of Potential Industrial Sources of Groundwater Contamination in the Tucson International Airport Area

Facility	Years Operated	Number of Employees	TCE Use	Chromium Use	Remarks
Consolidated Aircraft	4–6	Unknown	Possible, unquantified	Possible, unquantified	Nature of activities appears to make widespread Cr and TCE use unlikely
Grand Central Aircraft	4	Up to 4700; varied	1200–4800 gal/yr (est.)	129 lb/yr (est.) from TURCO products; plating known but use of Cr unquantified	Widespread use of additional chemicals; probable source of groundwater contamination in early 1950s
Douglas Aircraft	4	1500	None, according to former workers	None, according to former workers	Despite former employee accounts, nature of activities activities makes some TCE use likely, although quantities were probably relatively small
Tucson Aviation Center	15	Unknown, possibly several hundred	300 gallons documented; est. waste = 192 gal/yr	Negligible use documented in §3007 letter responses	Aggregate of numerous small businesses
Burr-Brown	20		267–10,000 kg/yr	Up to 132 lb/yr	Waste disposal practices highly suspect
West-Cap of Arizona	22	Unknown	2000 gal/yr	None documented	Poor accounting of waste TCE; improper disposal
Hughes Aircraft Co./Air Force Plant 44	34	Up to 6000	3000 gal/yr (est.)	6800 lb/yr (1959 est.)	Documented poor chemical handling and disposal practices; multiple potential source areas on-site

ered in this and previous investigations, the following conclusions were reached regarding possible *sources of groundwater contamination*:

1. *Air Force Plant #44* is an acknowledged source of chromium and TCE contamination. Use of chromium at the plant was the largest documented in the area; TCE use estimates were among the largest. The duration of large-scale use of chromium and TCE was longer at this plant than at any other potential source, including aggregate use at the Tucson airport hangar area. The site has had a number of potential sources, including *pits* in which spent chemical and sludges were disposed of and a wastewater discharge which was not retained on-site until 1961. Historic documents indicate **careless chemical handling** in areas where drainage systems allowed chemicals to flow directly to open washes, in some cases bypassing the plant's wastewater treatment systems. Historic analyses of *Tucson well SC-7* show that the **plants's effluent** was probably responsible for elevation of chromium levels in groundwater as early as *1958*. Analyses of soils and perched groundwater indicate that disposal pits and the historic wastewater were both probable means whereby contaminants, including TCE, entered the regional groundwater system. The available data appear to be consistent with the hypothesis that Air Force Plant #44 is the most significant source of groundwater contamination in the vicinity of the Tucson International Airport.
2. The **Grand Central Aircraft Company** almost certainly caused the contamination of local wells through the improper disposal of **wastewater**. While this wastewater probably contained chromium and TCE, these were probably relatively minor constituents compared to other chemicals known to have been supplied to the plant by *TURCO Products, Inc*. These other chemicals are known to be capable of causing groundwater pollution but are not now found in local groundwater. **Chromium plating** took place to some degree at Grand Central, although no estimates of usage were discovered, nor was a means of disposal for **plating wastes** generated. Use of TURCO products may have accounted for approximately 130 pounds of **chromium** per year. TCE use at Grand Central may have been as great as 4,800 gallons per year. Waste TCE from Grand Central was disposed of primarily at the **Tucson Airport Authority landfill**, which may have received as much as 2,400 gallons of TCE per year according to reliable witnesses. While activity at Grand Central was intense and corresponding estimates of TCE use were large, the duration of such activity at the plant was brief, lasting for probably little more than two years of the company's four-year tenancy. Primarily for this reason, Grand Central's potential for contribution of TCE to groundwater, although highly likely, appears to have been much smaller than that of Air Force Plant #44. Indications of chromium use at Grand Central comparable to that which took place at Air Force Plant #44 have not been discovered.
3. Information on the activities of **Consolidated Aircraft** at the hangars is very limited, but allows for the possibility that this facility contributed to groundwater pollution. Neither the existence nor the improper disposal of chromium or TCE have been reliably demonstrated at Consolidated. This facility's role as a new-aircraft modification center engaged primarily in assembly and installation would seem to preclude use of TCE or chromium on as large a scale as at Air Force Plant #44 or Grand Central Aircraft.

4. **Douglas Aircraft** does not appear to have been a significant source of groundwater contamination during the tenancy at the hangars based on an analysis of work performed there and information supplied by former employees. No TCE or chromium use or disposal was documented at Douglas, although the nature of work performed there allows for the possibility that some use of TCE or similar solvent occurred.
5. The **U.S. Air Force** occupied the Tucson Airport Authority hangars for approximately one month in 1968–69, and reportedly disposed of hundreds of gallons of liquids, including JP-4 jet fuel and TCE, **in the desert south of the hangars**. Subsequent EPA intermediate depth soil borings at the reported disposal sites failed to show evidence of vadose zone contamination, however.
6. Recent operators at the Tucson Airport Authority hangars include **small businesses** such as aircraft modification/repair companies of a type known to generate waste solvents. The largest reported use of TCE by any of these businesses is approximately 50 gallons per year. Using figures gathered for similar Maricopa County businesses, the aggregate waste solvent generated by the aircraft modification firms currently in residence at the hangars was estimated at approximately 200 gallons per year. While such figures leave open the possibility of groundwater contamination emanating from small businesses at the hangars, they are small in comparison with figures derived for other potential sources. The contaminant contribution of recent activities (that is, post-1970) thus appears to be minor.
7. Intermediate depth **soil sampling**[4] performed by EPA in the vicinity of the airport hangars failed to find evidence of **vadose zone**[5] contamination. Shallow soil samples taken near the hangars by ADHS did indicate disposal there of TCE and, in one instance, chromium. High levels of DCE, a volatile compound, in one of these samples may be indicative of recent disposal activities. Certain aspects of these sampling results, in particular the predominance of DCE over TCE, do not correspond well to conditions in underlying groundwater. Neither did the high chromium level found in one sample near the entrance to the hangars correspond well to known disposal practices there. In general, the presence of contaminants in shallow soil samples could not be conclusively traced to individual tenants or specific disposal activities at the hangars.
8. Three landfills were evaluated as to their probable groundwater pollution potential. Of these, only the old **Tucson Airport Authority landfill** appears to have received hazardous materials. TCE from Grand Central was dumped in the TAA landfill, along with other waste chemicals. While this is the only known dumping of TCE there, the possibility that other dumping of hazardous materials took place over the landfill's long history cannot be eliminated. Deep **soil borings**[6] contained TCE, possibly indicative of historic dumping. The TAA landfill appears to have had the potential to contribute TCE to local groundwater.
9. **Burr-Brown Research Corporation** is a highly probable source of local groundwater contamination located to the east of the main plume. This con-

[4] *Soil sampling* refers to the collection of soil samples for the primary purpose of investigating for the existence of contaminants.

[5] The *vadose zone* is the portion of the soil column that normally contains air as well as water.

[6] *Soil borings* are borings made primarily to obtain information regarding the conditions and properties of the soil, especially the degree of their contamination.

clusion is based on documented TCE use, poor disposal practices, and the reports of witnesses indicating the presence of an abandoned well on-site and the possibility of disposal there. More monitor wells are needed in the area to define Burr-Brown's contribution and differentiate it from possible contributions from its neighbor to the south, West-Cap Arizona.

10. **West-Cap Arizona** is a possible source of local groundwater contamination located to the east of the main plume. This conclusion is based on documented TCE use and the high probability of long term inadequate disposal practices. A **monitor well, SF-3,**[7] located down-gradient of part of the facility, may not be situated properly to monitor contamination from all portions of the plant. More on-site investigation and additional monitor wells are needed to more completely assess West-Cap's pollution potential.
11. The **Arizona Air National Guard** facility located at the northern edge of the airport is a probable source of local groundwater pollution east of the main plume. This conclusion is based on circumstantial evidence, largely the Guard's location relative to the known extent of contamination, the documentation of at least some TCE use, and the presence of possible pollution sources at the oil–water separators. It is not yet clear how activities at the Guard facility specifically relate to observed contamination. An ongoing **Installation Restoration Program** study of hazardous waste generated at the facility should allow better understanding of this relationship.
12. The **abandoned fire-drill areas** located **near runway 3** were in use from 1964 until sometime in the 1970's. While these areas received primarily JP4 jet fuel, they also received waste materials, possibly including TCE, from the Arizona Air National Guard. Intermediate-depth soil sampling performed by EPA at these sites failed to confirm vadose zone contamination emanating from them. The fire-drill areas currently in use are located in the southeastern portion of the airport north of runway 29. **Shallow soil sampling** here revealed high concentrations of a range of contaminants, including TCE. **Deep soil borings** contained traces of TCE and higher levels of toluene and benzene, indicating that downward migration of contaminants from this source has taken place. The fire-drill areas currently in use are a potential source of groundwater contamination. No local wells exist to determine the extent of this contamination, however.
13. The possibility that surreptitious dumping of TCE or chromium at as yet undiscovered locations near the airport contributed to groundwater pollution was not addressed in this investigation. The location and amounts of contaminants in the local groundwater system appear to be explainable on the basis of the activities previously discussed.

To summarize, there appear to have been two important sources of groundwater pollution which contributed to the main contaminant plume near the Tucson International Airport. Air Force Plant #44 appears to have been the more significant of these, while the activities of the Grand Central Aircraft Company appear to have been less impor-

[7] A *monitor well* is one that has been constructed primarily to sample groundwater for contamination and to measure groundwater elevations. It is normally sampled on a regular basis, such as every three months.

tant. The Burr-Brown Corporation, West-Cap Arizona, and the Arizona Air National Guard are probably responsible for two smaller contaminant plumes located east of the main plume. Other potential sources were considered to be less significant, if indeed they were sources as all, or could not be fully evaluated on the basis of available data.

1.2 AREAL EXTENT OF A MODEL

Let us now consider the first of the model construction steps outlined above, determination of the **areal extent of a model**. The areal extent of a model must be such as to

1. **Incorporate all locations where model heads are expected to change in response to stresses** imposed on the model. For example, when pumping at one or more wells to create a **cone of depression**,[8] the model should be large enough to include all areas where a decline in water level can be expected to be significant. By *significant* we mean declines that are likely to impact the overall groundwater flow and transport in the area of interest. Since such water-level changes normally are determined via the model output, such an area nearly always can only be approximated;
2. **Incorporate the area of interest to the client**. As an example, the client may be interested in seeing the simulated water levels or flow directions over an area larger than the area where water-level changes are to be expected. In order to have this flow information available for output, the applicable area should be encompassed within the perimeter of the model;
3. Result in a model that is **consistent with available computational capabilities**. In other words, if a personal computer is the largest computer platform available, the model size should be no greater than that for which an acceptable turnaround time can be realized on the personal computer platform. It is inappropriate for a groundwater professional engaged in modeling to remain idle for extended periods of time waiting for modeling results because of computational limitations.
4. To the degree possible, coincide with an **area defined by distinct and easily evaluated** *hydrological boundary conditions*.

Tucson Example

Via this field example we demonstrate, step by step, throughout the remainder of the book, how to utilize the GMA to model groundwater flow and transport. As noted in Part 3, the GMA is composed of the Argus ONE GIM system and the *PTC* groundwater flow and transport model. The GUI that interfaces these two programs is a plug-in extension (**PIE**). The *PTC* GUI-PIE installs its menu commands in Argus's

[8] A *cone of depression* is the area around a well whereat the water levels drop in response to pumping at the well.

PIEs menu. Thereafter it acts as a control panel for the creation of new *PTC* projects, the editing of control parameters of existing *PTC* projects, the execution of *PTC*, and postprocessing of the *PTC* output.

Since this is the first time we have faced the prospect of actually accessing the GMA environment, it seems appropriate to provide an abbreviated overview of the software that is used. More detail on each step may be found in subsequent sections of this book.

The steps involved in the creation, execution, and evaluation of a groundwater-flow model are the following:

1. After starting Argus ONE, the Argus ONE window appears and the user begins model development by selecting *New PTC Project...* from the *PIEs* menu.
2. A dialog box presenting the choices such as mesh type and number of geological (formations) layers then appears. The choice that is made causes the PIE to structure the kinds of geospatial coverages (information and data layers) required for a *PTC* simulation and automatically makes them available to the user for data entry and manipulation.
3. Next, the user may enter simulation-control parameters (those that are not spatially dependent, such as time-step size) into an interactive, tabbed-dialog box that appears on the computer screen. Upon completing data entry or editing of these values, the user closes the dialog and returns to the Argus ONE window.
4. The user should then modify the default information in any geospatial information layer by manually drawing closed or open contours or points to represent the desired spatial distributions of hydrogeologic and hydrologic parameters, fluid sources and sinks, and boundary conditions. One must also specify a desired finite-element mesh density. As an alternative to drawing, any of these spatial distributions may be imported directly from other applications that can generate either simple text files, DXF (Autocad format) files, or Shape (ArcView format) files.
5. The user now requests that Argus ONE create the finite-element mesh. Before proceeding to run *PTC*, the user may modify any of the spatial or nonspatial information already input.
6. The user then selects the *PTC Mesh* layer, and from the *PIEs* menu proceeds to "export" the geospatial and nonspatial information by selecting the *Run PTC* option. At this point Argus ONE writes out the standard input-data files for *PTC* to the directory selected, and runs the *PTC* simulation. When the simulation is complete, the user may choose to plot any of the simulation results within Argus ONE in a postprocessing step provided by the *PTC* PIE.

The power of the GMA approach in hydrogeologic hypothesis testing and practical modeling should be apparent at this point to any experienced modeler. In particular, it is straightforward to return at any point in the model formulation to any type of spatial information already input to the *PTC*–Argus ONE environment. It is also easy to make major modifications to this information or to the finite-element mesh, to export once again from Argus ONE, to run *PTC*, and finally, to graphically evaluate the results of new simulations. Each cycle of changing information, running *PTC*, and inspecting the results can take as little as a few minutes.

Let us now return to the practical aspects of setting up a model. To access the interface, activate Argus ONE. An existing project can now be selected by clicking on *File* and then *Open*. From the resulting dialog box, one can then select an existing *.mmb* file produced during an earlier investigation (Figure 1.3).

Alternatively, if a new project is to be considered, one can click on the *PIEs* menu and select *New PTC Project....* Plug-in extensions (PIEs), as noted earlier, are functions or groups of functions that add capabilities to the basic structure of Argus ONE. The *PTC* PIE contains the following functions:

1. A function to create an Argus ONE project for *PTC*. This function is executed by selecting *PIEs | New PTC Project....*
2. A function to edit the project information, executed by selecting *PIEs | Edit Project Info....*
3. A function to run *PTC*, executed by selecting *PIEs | Run PTC*.

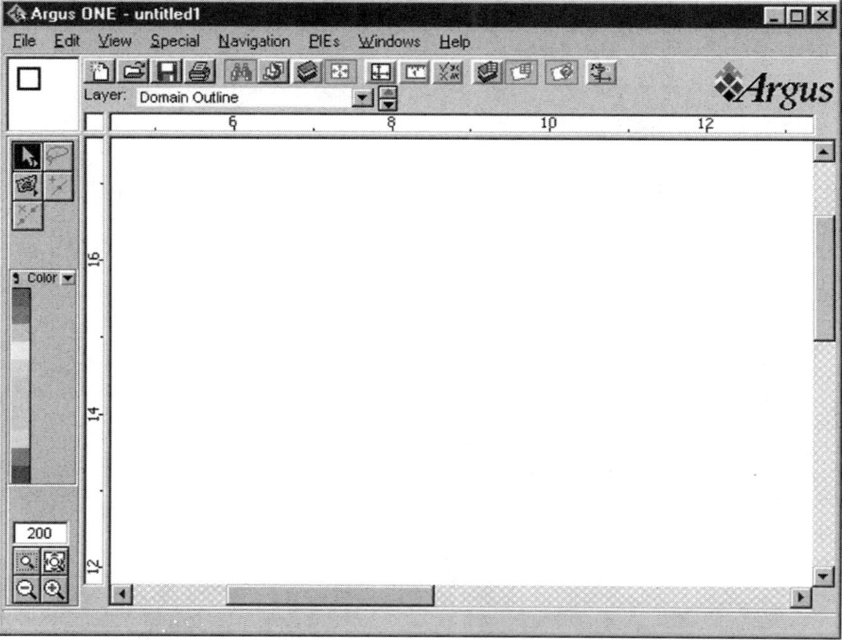

FIGURE 1.3. *The first step in setting up the PTC model using Argus ONE is to establish that the model to be used is PTC. This is accomplished by selecting New PTC Project... from the PIEs menu.*

The first two of these functions are always available in the *PIE* menu. The third is available in the *PIE* menu when the *PTC Mesh* layer is active. It also contains several functions that are hidden to the user. This group of functions is created as a *library* that is linked to Argus ONE whenever the user starts the program. You can see a list of all the libraries that Argus ONE loads during startup in the first window it displays. All of these libraries are linked when execution starts. Traditional linking of libraries is made when a program is compiled, so this type of linking that does not require one to recompile is called *dynamical linking*; so our *PTC* PIE is a dynamic-link library (DLL). A *.DLL* file contains one or more functions compiled, linked, and stored separately from the processes that use them.

After selecting the *New PTC Project* option, a *PTC Configuration* window will open, as shown in Figure 1.4. The first order of business is to identify the project by giving it a convenient, descriptive name. *PTC Project* is the project title provided in the *Project title* dialog box shown in Figure 1.4. One must also indicate whether the water-table feature of *PTC* will be implemented. In other words, should the uppermost layer of the model be treated as an unconfined aquifer? By checking the *Use water table* box, the water-table feature is invoked. In this event, it is necessary to identify the number of iterations to be used in solving the free-surface problem *(Number of iterations for water table* dialog box*)* and the criterion to be used to indicate convergence of the solution *(Convergence criterion* dialog box*)*.

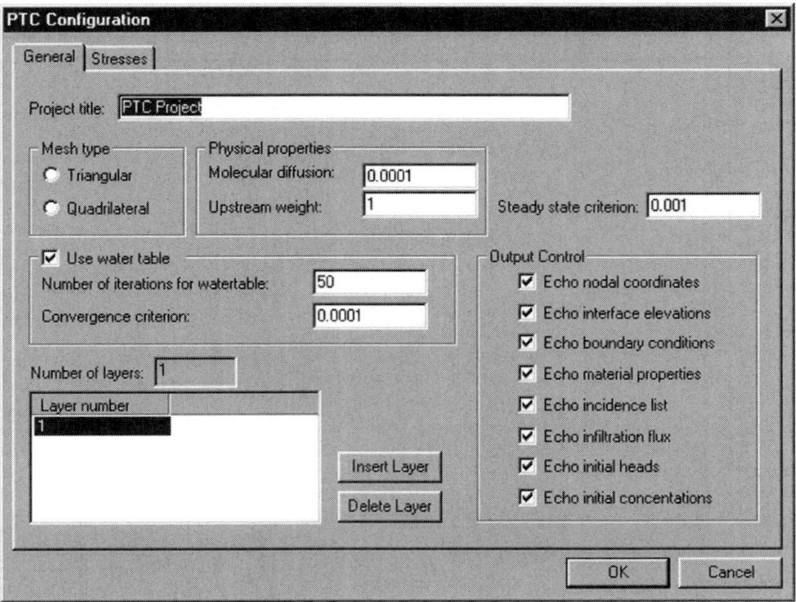

FIGURE 1.4. *PTC Configuration dialog box opened by selection of New PTC Project.*

It is now necessary to define the number of layers (vertical discretization) to be used in the model. Because we have decided to have one layer in our model, a default value of 1 appears in the dialog box identified with *Layer number*. Additional layers can be added by clicking the *Insert Layer* button. The paradigm for adding layers is as follows: If you have one layer in the listbox and it is highlighted, pressing the *Insert* button will insert a second layer as layer 1. The layer you originally highlighted will now be layer 2. Similarly, layers can be deleted by highlighting the layer to be removed and then clicking the *Delete Layer* button.

Under the heading *Output Control* are eight boxes that control the form of the *PTC* output. The purpose of the "echo" format is to permit the user to confirm the information being used as input to *PTC*. Normally, this information is accessed only if a problem arises in execution of the model proposed. We will return later to complete the items in this *General* dialog box.

Next select the *Stresses* tab in the *PTC Configuration* window. The window shown in Figure 1.5 appears. Under the heading *General control*, designate whether the model is to simulate flow by checking the box identified with *Do flow* or flow and transport by checking *Do transport*. If velocity calculations are required, click on *Do velocity*. The option of *Use memory* should always be implemented. A mass balance calculation is optional and, if desired, can be activated by checking *Do mass balance*.

The *Graphics filenames* text box is used to specify the names of the files to be assigned to the graphics output. In Section 1.17 we describe the details of how the output is formatted. For now it will suffice to place a convenient name in the text box. Here we have chosen the numeral 1. At this time we will not provide the remaining information requested in this window, but rather, revisit and complete it later as we develop the appropriate background.

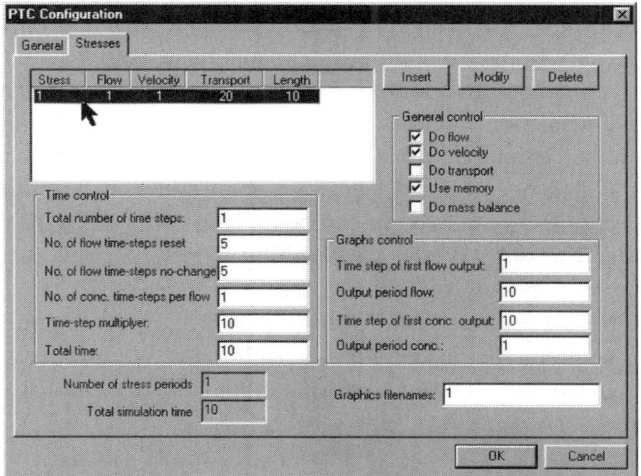

FIGURE 1.5. Multiple stress period information is provided via this window along with various other project information.

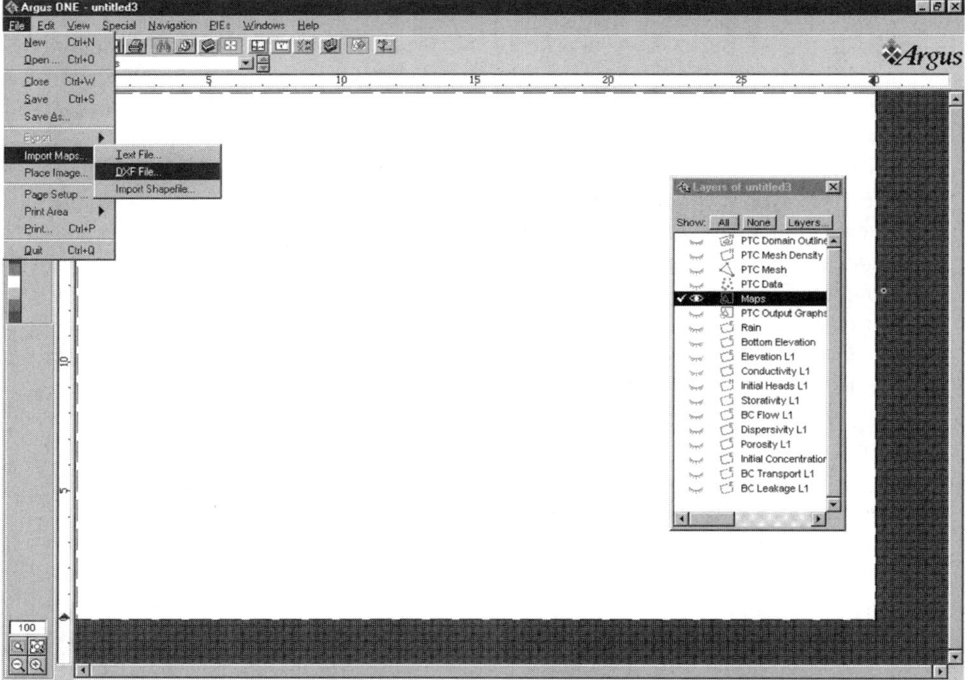

FIGURE 1.6. *Window used to import a .dxf file that contains the information required to produce a base map in the Argus ONE environment.*

Project Boundaries Let us now consider the specification of the modeling-project boundaries. As a first step in defining the **model boundaries**[9] and to provide a visual reference point for evaluating data input and output from a groundwater model, it is necessary to have a computer-readable **base map**.[10] The base map provides the geographical information needed to accurately position hydrogeological and hydrogeochemical information important in defining the groundwater model. For the above-defined problem, a base map was generated by first scanning an existing map of the area of interest. Using a graphics editor, a simplified, computer-readable version of this complex map can be created using standard drawing tools. In the case of the Tucson example, the resulting map was saved as a *.dxf* file and imported directly into the Argus ONE interface.

The Argus ONE screen that permits this import is shown in Figure 1.6. Consider first the pull-down menu that appears on the right-hand side of this window. If this menu does not appear, click on the *Layers* icon, which is the third from the right in

[9] *Model boundaries* define the perimeter of the model in three space dimensions. They have specific mathematical definitions, which we address in a later section.

[10] A *base map* presents the principal geographical features identified with a site. Roads, existing wells, buildings, and so on, are normally recorded on such a map.

the toolbar at the top of the project window. In this *Layers of model_new...* menu are found the various *layers* that will be used to input data to *PTC*. Because we have one layer, this menu only requests information for layer 1. Keep in mind that the layers indicated here are model or project layers, not geological or aquifer layers. There may be several project layers for each aquifer layer.

The corresponding menu for a two-layer system is shown in Figure 1.7. Note that in this figure there are additional layers associated with the second layer, designated as L2. The new layer is by convention higher (nearer the surface) than that with the lower number, that is, L1.

In Figure 1.7, we also see that there are a series of *eyes* located on the left-hand side of the *Layers of untitled2...* drop-down menu. If an eye is open, the information in the layer to the right of the eye is visible on the project window. In this instance we have yet to provide information, so the window is clear of any image.

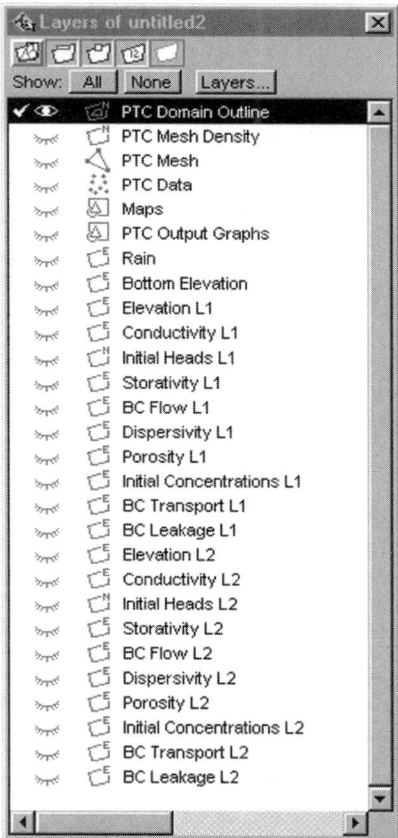

FIGURE 1.7. *Window illustrating the layer sequence for a two-layer system. Note the L2 designation associated with information for the second (upper) layer.*

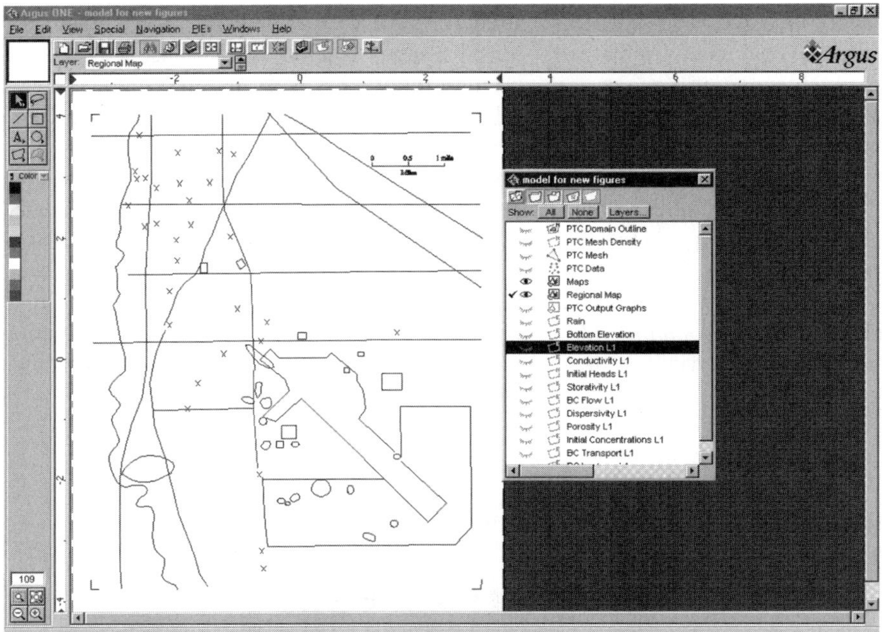

FIGURE 1.8. *Location map of the Tucson airport area generated by the Argus ONE environment from a .dxf file input. The labels for the wells and contaminant sources have been added.*

On the left-hand side of Figure 1.6 are two pull-down menus available when a *Maps* layer is chosen (by clicking to the left of the eye) from the *Layers* window. The window overlap in this figure indicates that one first selects *Import Maps*, which takes one to the window in which the *DXF File* option is selected. Clicking on this option, one obtains a dialog box wherein the appropriate *.dxf* file can be designated. For the Tucson example, the base map is as shown in Figure 1.8.[11] If a graphics editor capable of generating a *.dxf* file is not available, there is an alterative strategy. Make the *Maps* layer active and click on *File*. Select *Place image*. . . . The file type you can now import can be a bitmap such as a .gif file. Note that many project layers can be visible at the same time ("open eye") but only one is active at any given time.

Scaling the Map The *project figure*, for example the base map, now located on the screen is defined in terms of screen coordinates. **Screen coordinates** are those identified with the scales located on the top and left edges of the Argus ONE project window. A scaling relationship is needed to transform the figure to **field coordinates**. The field coordinates are those that correctly define the various geographical elements represented in the *Maps* layer and other layers to be defined later.

[11]The base map was created through tracing of the original hard-copy figure by students participating in a modeling course taught at the University of Vermont.

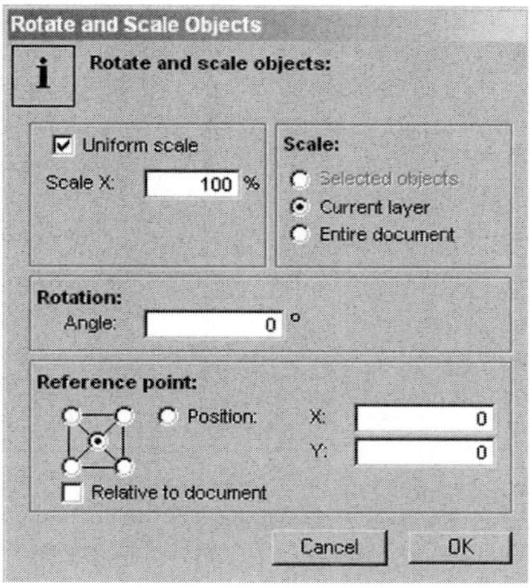

FIGURE 1.9. Rotate and scale objects dialog box for adjusting the scale of the map to the scale of the computer screen.

To define the coordinate transformation, make the *Maps* layer active.[12] Use the *zoom* magnifier to expand the image by a factor of 2 to allow you to see more clearly the scale that appears on the map. Place the cursor on the left side of the scale and record the location; it should be 1.14 cm. Place the cursor on the right-hand side of the scale and record the location; it should be 2.29 cm. Subtraction gives a value of 1.15 cm. Thus 1.15 cm on the screen represents 1 mile according to the scale on this base map. It is convenient, for reasons that will become obvious momentarily, to have the map scale represented by integer values. In other words, it would be helpful, in this example, to have 1 cm in *screen units* represent 1 mile in *map units*.

To realize this goal, click on *Special* and then select *Rotate and Scale* from the menu bar. A window such as that shown in Figure 1.9 appears. Now we wish to adjust the scale of the map so that 1 screen unit equals 1 mile. To do this, we divide the screen unit length 1.0 by the length of the scale, namely 1.15, to obtain 0.87. Thus we wish to reduce the scale to 87% of the original to achieve a one-to-one correspondence between the screen unit and the 1-mile length scale. To be sure that the entire drawing is modified, click the *Entire document* button.

Replacing the value of 100% in this dialog box by 87% and clicking *OK* achieves the desired transformation. The redrawn map will now have 1 screen unit equal to 1 mile on the map. The de facto impact of this decision and subsequent transformation is that **all parameters that are used in the model must now have length units of miles**.

At this point the choice of time units is undefined. Once a parameter that uses time is specified, the time units used for that parameter will define the time scale. In this example, days will be used for the unit of time.

[12]Although we have suggested the *Maps* layer for concreteness in discussion, other layers will also work provided that the layer is of type *Map* indicated by the associated icon to the left of the name.

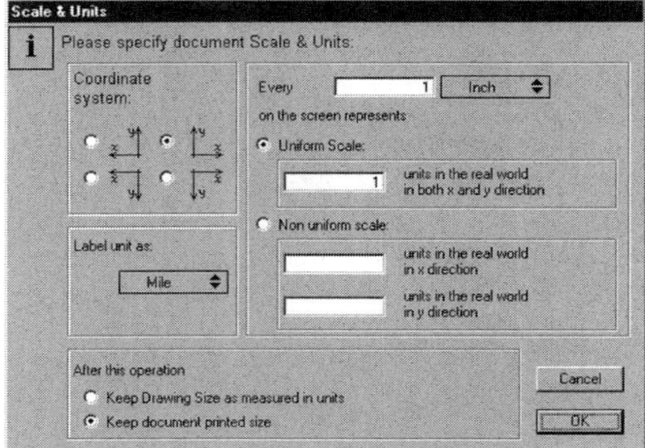

FIGURE 1.10. *Second of two windows utilized to establish the transformation between the screen coordinates and the actual field coordinates.*

If you would like to have the pointer location identified with units of miles, the true field scale, rather than centimeters, select the *Special* option and then the option *Scale & Units*. The *Scale & Units dialog box* will appear (Figure 1.10). Under the box labeled *Label unit as*, activate the arrow to select the units you wish to have appear associated with the pointer location on the Argus ONE screen, then click on *OK*. Your system is now scaled such that 1 screen unit, in this case 1 cm, equals 1 mile on the map.

If, for some reason, you would like to have 1 cm in length on the computer screen represent more or less than 1 screen unit (in other words, you would like to have the ruler larger or smaller), you can accomplish this by changing the values in the two boxes in the upper right-hand corner of this window. For example, if you change the *units in the real-world* value to 0.5, and click on *OK*, the redrawn map will be twice as large, although the actual scaling has not changed. In other words, 1 screen unit will still represent 1 mile, but a 1-cm length on the screen will represent 0.5 mile. As is evident from Figure.1.10, additional options are available within the Argus ONE environment and the reader is referred to the Argus ONE documentation for additional insight into these possibilities.

Finally, select the *Special* option and click on *Drawing Size*. The window that appears is shown in Figure 1.11. The drawing defined in this dialog box is a standard 8.5-by 11.0-inch page with the coordinates located in the lower-left hand corner of the page. However, on occasion, one may wish to select a subset of a large map as the project map. For example, if the map containing the study area is 4 ft by 4 ft in size and you wish to consider only a 1 ft by 1 ft square in the middle, you can specify this in the *Drawing Size* dialog box. The *Horizontal Extent* and *Vertical Extents* would be 1.0 ft and the *Horizontal Origin* and *Vertical Origin* would be located at (1.5,1.5).

Selection of the **minimum size** of the model depends, as mentioned earlier, on the needs of the client, the available computational capability, and the anticipated physical response of the system. Because we anticipate the need to reproduce the

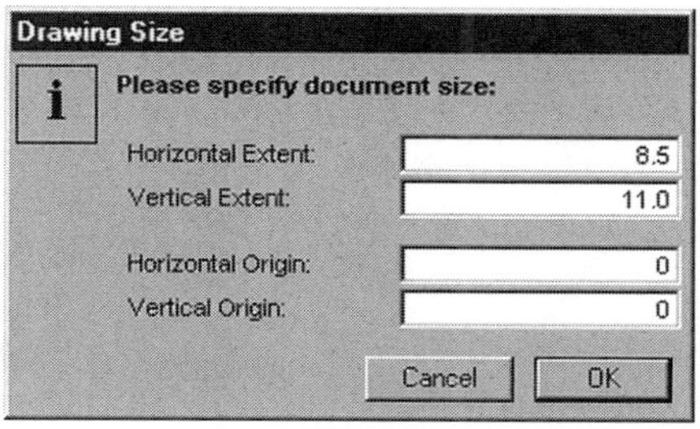

Drawing Size

Please specify document size:

Horizontal Extent: 8.5

Vertical Extent: 11.0

Horizontal Origin: 0

Vertical Origin: 0

Cancel OK

FIGURE 1.11. *Drawing size dialog box that relates the size of the drawing to the screen units. In this example, it is assumed the origin is in the lower left-hand corner of the scanned map and that the paper is standard 8.5 in. by 11.0 in.*

contaminant plume at the Tucson site[13] with our model, **the model dimensions must be at least as large as, and encompass, the plume area**. Figure 1.2 illustrates the observed plume geometry. If we expect to forecast the future behavior of the plume, a task not anticipated for this particular model, **the model boundaries would normally encompass the anticipated maximum size of the plume over the period of analysis**.

The second issue of importance in this model is the **anticipated water-level response**. In general, the water-level changes created by pumping wells will propagate rapidly over long distances, often to areas beyond the current or anticipated contaminant-plume boundary. Thus the need to respond to anticipated water-level responses will often result in a model of larger size than that required solely for simulation of an associated contaminant plume. In our Tucson example we observe that there are numerous high-capacity water-supply wells in the neighborhood of the plume. The cone of depression (area of influence) of these wells will extend beyond the current or anticipated plume perimeter. Thus the boundaries of the model must be extended beyond those that would be required solely for contaminant-transport simulation.

Defining the Model Geometry Based on an analysis of the issues noted above, boundaries for the Tucson model domain were selected. These boundaries are used in Section 1.5.2 to define boundary conditions on the model. To input this information, one proceeds in one of two ways. One alternative is to define the boundaries by specifying the coordinates of points along the boundary. A second approach is to draw the boundary on computer screen using the mouse.

In Figure 1.12 is illustrated the protocol used to input the model-domain boundary. While in the *PTC Domain Outline* layer of the *Layers* floater, one uses the *geographic tool*[14] shown active in the toolset located in the upper left-hand corner of the window

[13]The term *plume* normally refers to that portion of the groundwater system contaminated to beyond a specified threshold concentration by one or more sources in the region of interest. It is contiguous.

[14]By resting the cursor over a tool, its definition will be provided.

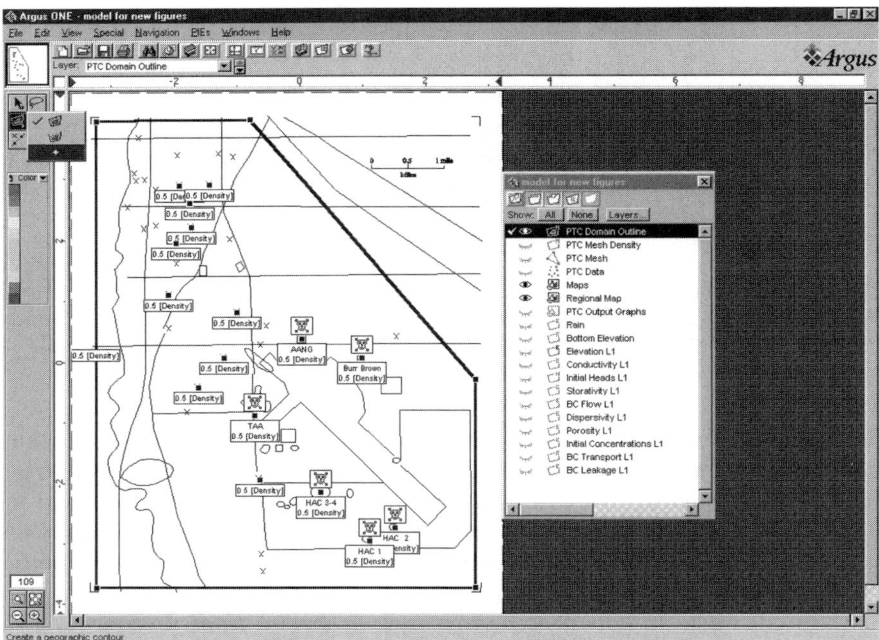

FIGURE 1.12. *Definition of model domain boundary using the mouse-definition option. Note that not only the external boundary, but also the locations of point boundaries, such as contaminant sources have been defined.*

in Figure 1.12. Having located the cursor, which in this instance will appear as a plus sign, at some location on the proposed boundary, one clicks the mouse to begin the boundary definition. By moving the cursor from point to point along the boundary, and clicking the mouse at each point, a polygonal representation of the boundary is generated. At the final boundary location, one executes a double mouse click and the boundary is defined. The boundary in Figure 1.12 is represented by the bold line that forms a five-sided polygon.

Defining the Location of Point Boundaries Also defined in Figure 1.12 are the points that will be used later to define the **locations of point sources and sinks**. In our case these will consist of pumping and discharge wells as well as contaminant source locations. The tool used for this is the *Geographic* tool, which, as noted, is highlighted in Figure 1.12. By clicking and holding on this tool, the menu of icons shown in Figure 1.12 will appear. The three icons correspond to closed, open, and point contours. Select the *Point* tool (plus sign) in the submenu illustrated in Figure 1.12. The *Geographic* tool is now replaced by the *Point* tool. Click on the point tool and drag the cursor to the point in the domain where you intend to place a point boundary condition, for example a well, and click. The dialog box shown in Figure 1.13 will appear. A similar box requesting domain outline density information

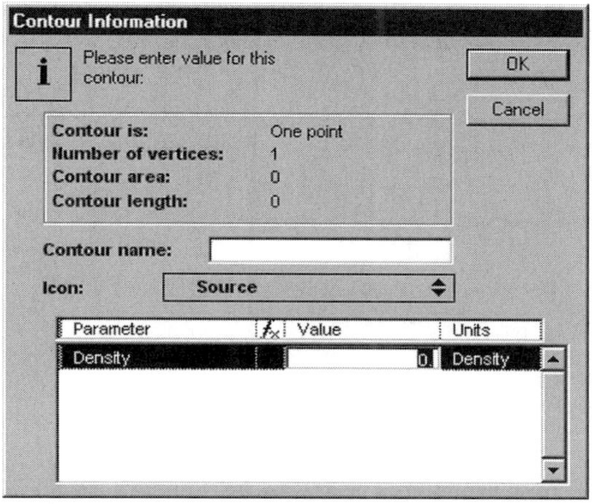

FIGURE 1.13. *Dialog box used to define point conditions. Further definition of the point condition will be discussed later in the text.*

is created if one double-clicks the domain outline. We discuss later the information to be provided in these dialog boxes.

Now double-click the spin box identified as *Source*. The submenu illustrated in Figure 1.14 results. An appropriate icon can be attached to the point condition identified in the domain by moving the cursor to the desired icon location. One can now return the cursor to the domain to select the next point-condition location. Although placement of the point conditions illustrated in Figure 1.13 may seem premature at this point, the objective of selecting these locations at this juncture is to provide a placeholder for information to be provided later in selecting boundary conditions.

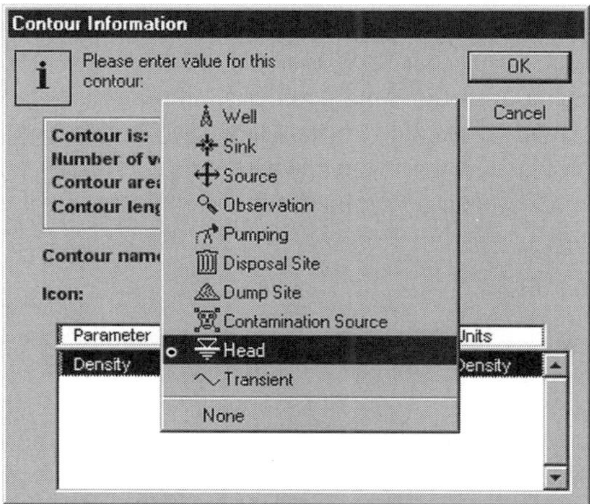

FIGURE 1.14. *Submenu used to provide an identifying icon for point conditions.*

TABLE 1.2. Format to Be Used to Input Boundary Information Using the *Text* Option[a]

Format for Closed Boundaries	
Number of points of the outline +1[b]	Density
x coordinate	y coordinate
Format for Point Boundaries	
1	Density
x coordinate	y coordinate

[a]The parameter *density* is considered in Section 1.15.

[b]Note that the first point should be repeated at the end to force the outline to be a closed contour.

Note that a more comprehensive description of the use of the various tools can be found in the Argus ONE documentation.

An alternative strategy for defining the domain boundaries is to use a text file with the extension *.exp*. The format is shown in Table 1.2. The x and y-coordinates are listed columnwise until the last point, which is repeated (clicked a second time) to force the outline to be closed.

1.3 HYDROLOGICAL BOUNDARIES TO THE MODEL

Although not essential, it is advantageous to use well-defined and distinct **hydrological features as boundaries** for the model. This is due to the fact that the model is separated from the rest of the world by what is specified along the model boundaries. Therefore, a feature that can be defined quantitatively, such as the water level of a surface-water body, is a desirable hydrological boundary for a model.

Alternatively, **geological structures** or **changes in rock type** can be used to define model boundaries, rather than distinct hydrological features. Such an approach is appropriate when rock formations serve to act as impermeable barriers. For example, the interface between a gravel or sand geological unit and a relatively impermeable bedrock unit can provide a suitable impermeable barrier for a model that simulates the behavior of groundwater in gravel and sand units.

Features that typically make suitable boundaries to the areal extent of a model include the following:

1. Lakes
2. Large ponds
3. Rivers
4. Estuaries
5. Ocean shorelines
6. Unlined canals
7. Human-made reservoirs

8. Impermeable barriers due to changes in geological materials
9. Impermeable human-made barriers such as walls
10. Human-made sink terms such as drains
11. Human-made source terms such as infiltration galleries

Tucson Example

Because in an arid climate there is a general lack of surface-water bodies connected hydrodynamically to an aquifer, it is often difficult in such circumstances to utilize classical hydrological boundary conditions for a groundwater flow model. Such is the case for the Tucson model that we are using as an example.

The Santa Cruz River, shown in Figure 1.2, could play a role in the determination of groundwater-flow if it were a perennial stream hydraulically connected to the groundwater system. It was the opinion of the modeler in this instance that this was not the case, and therefore the Santa Cruz river was not considered as a hydrological boundary. Without this or other surface-water features, other strategies, which we discuss shortly, had to be brought to bear to define the necessary boundary conditions along the model perimeter.

1.4 COMPILATION OF GEOLOGICAL INFORMATION

Because of the close relationship between the hydrological properties of groundwater reservoirs and the geological characteristics of the materials that constitute the reservoir matrix, it is helpful at this point to focus on the nature and compilation of geological information. For example, the hydrological properties of materials normally depend on the geological environment in which they are created. **Clay sediments**, for instance, having been deposited in quiescent conditions, have a very small grain size and an associated very small pore size (although the porosity of clay can be quite high). As a result, there are substantial friction losses as groundwater moves through clay deposits. Consequently, clay is a low-permeability material (i.e., it has a **low hydraulic conductivity**).

The areal and vertical **distribution** of clay units generally coincides with ancient bodies of water where the energy environment was low. Such areas may be very large, such as the bottom of a large lake. They may also be quite small, such as a quiet pool along the bank of a river.

Knowledge of the physical environment in which a deposit was generated can give insights into whether it has a large areal extent. Such knowledge can also provide insight into whether the clay layer can be expected to be continuous or, as is often the case in the field, to have areas where it is missing due either to erosion after deposition or because of a lack of deposition altogether. Because of the important role that grain size plays in groundwater-flow and transport, let us discuss how it is measured.

Grain size is easily determined through the use of **sieves**. The classical sieve consists of a metal cylinder approximately 5 cm in height and approximately 20 cm

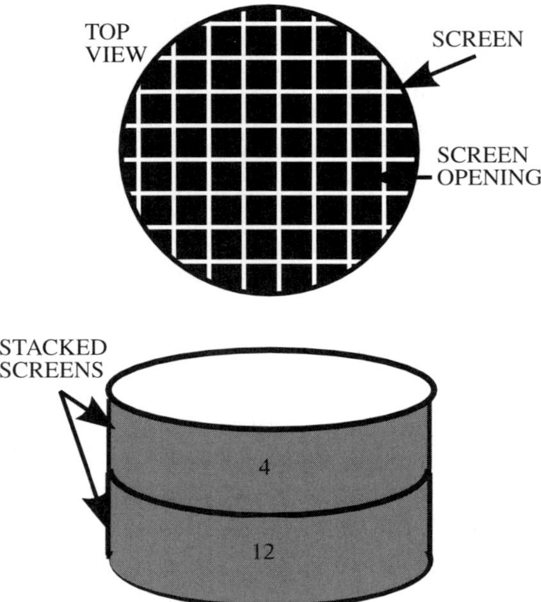

FIGURE 1.15. *Screens are stacked sequentially from the finest mesh at the bottom to the coarsest at the top.*

in diameter. It is open at one end and contains a metal screen at the other. Sieves are normally stacked with the sieve that has the smallest screen size opening, or **mesh size**, at the base of the stack. A pan is placed below the last sieve to collect those grains smaller than the smallest grain size captured by a sieve (Figure 1.15).

To sieve a sample of soil, a known weight of the soil is placed in the uppermost of a series of stacked sieves. This sieve is covered and a shaking apparatus is used to vibrate the column of sieves while they remain approximately vertical. The grains that are smaller than the opening in the top sieve eventually pass to the next-lower sieve. This sieve, in turn, retains those grains with a diameter larger than its mesh size and smaller than the mesh size of the sieve above. This process continues from one sieve to another until the grains retained in the container at the bottom of the column are smaller than the screen opening diameter of the sieve with the smallest mesh. The soil fraction resident in each sieve is then weighed and the results plotted.

Sieve sizes are designated in a number of different ways. Some sieves provide the sieve diameter in inches or millimeters. Others designate the sieve by its number, which has no obvious relationship to the mesh size. Typical sieve sizes are shown in Table 1.3.

Normally, material smaller than that captured by the No. 200 mesh screen (material with grain diameter less than 0.063 mm) is very difficult to screen further and is therefore analyzed via a "wet" method called the **hydrometer method**. This approach is used to separate silt from clay-sized particles and is based on the use of

TABLE 1.3. U.S. Standard Test Sieves (ASTM)

Sieve Designation		Nominal Sieve Opening	
Standard	Alternative	Inches	Millimeters
25.0 mm	1 in.	1	25.7
11.2 mm	7/16 in.	0.438	11.2
4.75 mm	No. 4	0.187	4.76
1.70 mm	No. 12	0.0661	1.68
0.075 mm	No. 200	0.0029	0.063

Source: Data from Anderson [2].

Stokes' law and a knowledge of the density of the water–soil suspension. Stokes' law is needed because it relates the velocity of a spherical particle falling through a fluid to its diameter and specific gravity. Why this is needed becomes evident in the following description of the procedure.

Imagine that we have already weighed the portion of a sieved sample that has been collected in the pan underlying the No. 200 mesh screen. We are now confronted with the task of determining how to find the size distribution of this sample of very fine grained material. The first step is to place the smaller-than-200 mesh screen sample in a graduated cylinder and add water until the resulting suspension is 1000 mL. Next we add a **deflocculating agent** so that the best possible particle dissociation is achieved. The resulting suspension is then agitated by covering the open end of the cylinder with one hand and inverting the cylinder several times.

For reasons that will become evident shortly, we next place a hydrometer in the solution and measure the density of the suspension that is found above a selected but arbitrary depth below its surface. We make this measurement at 0.5, 1.0, 2.0, 4.0, 8.0, 16.0, and so on, minutes after the suspension is created.[15]

Next we determine the weight of the sample composed of the various grain sizes smaller that the No. 200 mesh screen. From Stokes' law[16] we can calculate the size (in the sense of diameter) of the grain particle that is passing by our arbitrary plane at each of the times noted above: that is, 0.5, 1.0, and so on, minutes.

Since we know the size of the grains passing the plane of interest at these times, the outstanding question is: What weight of particles of a specified size has passed the arbitrary plane at each of these times? The answer lies in the fact that we know from our earlier measurement the density of the solution at the elevation of the specified plane at the measurement times. Thus we know the mass of soil particles in suspension above the plane at these times. The remainder of the soil particles must have passed by earlier and must therefore be larger than the size calculated to have been passing the plane at the times specified. If we keep track of the weight of par-

[15]Note that this selection of measurement times is not accepted universally. Different references in the literature recommend different measurement times.

[16]An important assumption that is made in using Stokes' law is that the grains are spherical. Although this may be appropriate for sand-sized particles, clay particles tend to be platelike, and some calibration of the procedure may be necessary.

TABLE 1.4. Experimental Results from an Hydrometer Method Experiment for Determining Fine-Grain-Size Distributions

Grain Size (mm)	Weight Smaller (g)	Percent Smaller
0.04	147.0	98
0.010	127.5	85
0.005	91.5	61
0.002	42.2	28
0.001	22.5	15

ticles in suspension at each time, we can determine the weight of particles smaller than the particle size calculated at each time. Table 1.4 provides the results of such an experiment conducted using a soil sample with a smaller-than-200 mesh screen size fraction of 150 g. It is because the density measurements are usually obtained using an hydrometer that the method is called the **hydrometer method**.

The information gained from a sieve analysis reveals more than just the range of grain sizes. It can also help to **classify the soil** as to its type (e.g., sand, silt, silty sand, etc.). In addition, it reveals the **degree of sorting** of the soil. Finally, the shape of the resulting **grain-size distribution curves** can also reveal information regarding the history of the soil. In Figure 1.16 the grain-size distribution curves for two soil samples are plotted. The grain size is plotted along the horizontal axis. On the vertical axis is plotted the percent weight finer than the indicated grain size. For example, the percent by weight of grains smaller than 0.01 mm in the clayey-sandy-silt sample is

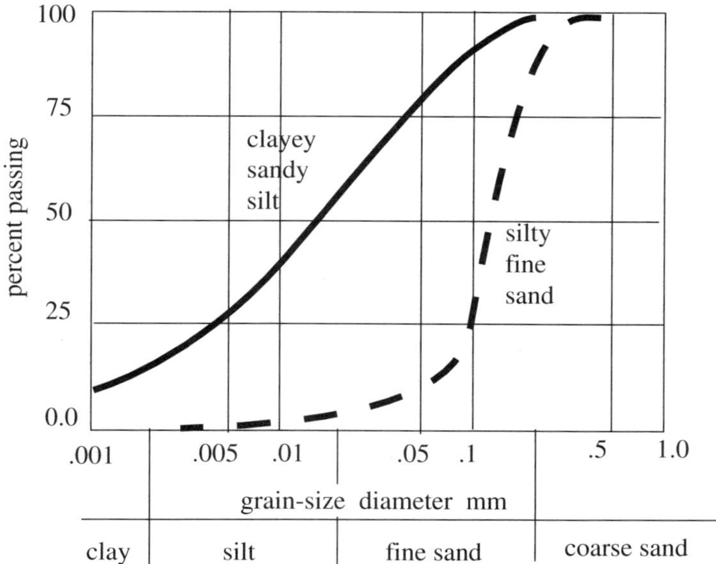

FIGURE 1.16. The grain-size distribution indicates the soil classification of a sample and its degree of gradation.

approximately 40 percent. Saying the same thing slightly differently, one could state that 40 percent of the grains, by weight, in the sample are smaller than 0.01 mm. Similarly, in the case of the silty-fine-sand sample, approximately 25% of the grains are smaller than 0.1 mm. It is clearly evident that the clayey-sandy-silt sample in this example is finer grained than the silty-fine-sand sample. In fact, by referring to the **soil classification** found beneath the distribution curve, it is evident how these samples received their classification.

However, there is more information that can be obtained from these curves. It is clear that the slope of the silty-fine-sand curve is much steeper than the slope of the clayey-sandy-silt curve. This tells us that the silty-fine-sand has a more **uniform size distribution**. In other words, the range of grain sizes in this sample is small relative to the clayey-sand-silt sample. The silty fine sand is considered to be better **sorted** or more poorly **graded** than the clayey sandy silt.

A measure has been developed to describe the range in grain sizes of a soil sample. It is called the **uniformity coefficient** and is defined as

$$C_u = \frac{D_{60}}{D_{10}} \tag{1.1}$$

where D_{60} refers to the grain size corresponding to the weight passing of 60%. In other words, 60% of the grains by weight are smaller than D_{60}. The denominator, designated as D_{10}, is also known as the **effective grain size**. The larger the value of the uniformity coefficient, the better graded (more poorly sorted) is the sample. In our example, the uniformity coefficient of the clayey-sandy-silt sample is

$$C_u = \frac{0.02}{0.001} = 20 \tag{1.2}$$

and that of the silty-fine-sand is

$$C_u = \frac{0.15}{0.05} = 3.0 \tag{1.3}$$

which confirms our earlier hypothesis that the clayey-sandy-silt is a better graded soil.[17]

Now that we have more insight into the concept of grain size and its measurement, let us consider now a few depositional environments of interest to the groundwater professional.

1.4.1 Unconsolidated Environments

Clastic Sedimentary Environment A **sedimentary environment** can be thought of as one in which all of the deposited material has been precipitated out of solution, derived locally from living organisms, or transported from elsewhere. **Clastic materials** are those that have been transported from elsewhere. Cobbles,

[17] See the *Earth Manual* (Bureau of Reclamation [1]) for more information regarding grain-size distributions and soil classification.

TABLE 1.5. Clastic Sedimentary Environments

	Clay	Silt	Sand	Gravel	Bioclastic
Beach	N	N	Y	Y	Y
Lake bottom	Y	Y	Y	N	Y
River	Y	Y	Y	Y	N
Delta	Y	Y	Y	Y	N
Ocean bottom	Y	Y	N	N	Y
Dune	N	N	Y	N	N
Loess	N	Y	N	N	N
Talus	N	N	Y	Y	N

boulders, sand, silt, and clay constitute materials that are normally found in a clastic sedimentary environment. Such materials are normally **transported by water**. However, this is not universally the case. There are materials that have been **moved by wind** to form **loess** and **dune** deposits. In addition, **gravitational forces** can move material loosened by erosion from highland areas to form a **talus slope**. Clastic materials may also be carried by ice and subsequently, by water and wind.

In general, when one encounters these granular materials, they can be identified roughly with the environments described in Table 1.5.

Precipitant Sedimentary Environment Under favorable conditions, **chemical precipitates** can form when the concentration of the chemical of interest reaches a saturation concentration. At this concentration, precipitation begins and the precipitate may then settle on the floor of a quiescent water body. The Great Salt Lake is an example; here halite (common table salt) is being precipitated. Calcium carbonate and calcium magnesium carbonate are also commonly encountered precipitates. In addition, it is possible to form a deposit through accumulation of the **remains of aquatic plants and animals**.

Glacial Environments Over the northern third of the United States and most of Canada, the surficial landforms are defined primarily by Pleistocene glaciation. At that time, massive **continental ice sheets** moved southward from polar regions to form ice thicknesses of more than a mile in some areas. The Missouri and Ohio Rivers roughly represent the margin of the ice sheet during its most southward advance.

In addition, in more mountainous areas, **valley glaciers** advanced to form large glaciated valleys. The glacial ice carved out surficial material in some areas and redeposited it in others. Much of the fertile soil of the midwest of the United States once resided in Canada and was transported via continental glaciation and subsequently by water and wind.

Because many of the important groundwater contamination problems are located in the northeastern United States, where the relevant aquifers are made up of materials deposited by the glaciers, it is important to understand the basic mechanisms at work and the resulting geological deposits.

Unstratified Deposits Material scraped from the earth's surface by the glacier and later deposited in various landforms is called **till**. Till is very unusual, inasmuch as it generally exhibits **no grain-size sorting**. In other words, grain-size analyses would show a wide distribution of grain sizes. This is because there is no mechanism for sorting the materials as there is in water or wind transport. As a consequence, in the same sample of till, one might find every grain size, from clay to boulder. The **composition of the till**, both mineralogically and in terms of grain size, depends on the source of material available to the glacier as it flows under gravity from its area of origin to its perimeter. For example, glacial ice moving through the Great Lakes basins tended to erode lake-bed materials, which contained considerable clay-sized material. As a result, this till is characterized by an abundance of clay-sized particles.

Till is categorized as ablation and basil. **Ablation till** is till deposited via transport to the surface of a glacier, whereupon, through melting, it forms a relatively unconsolidated deposit. **Basil till**, on the other hand, is deposited at the base of a glacier, not unlike butter being spread on bread by a knife. Basil till tends to be extremely hard, with characteristics similar to those of rock. Both kinds of till deposits are found over much of the area of the United States that has experienced glaciation.

Several characteristic landforms are generated by glacial deposition and erosion. **Moraines**, which come in a variety of forms, are the most common. One of these, **lateral moraines**, are formed where glacial ice contacts a valley wall. They appear as ridges of crushed rock and debris at the ice–rock contact. The related **medial moraines** are found where two glaciers come together. They are found along the line where the material that originally formed two of the lateral moraines, one from each of the two original glaciers, comes together. Thus the medial moraine is located on the interior of the combined ice flow rather than at the edges. Of course, there are still two lateral moraines on the combined glacier. Extensions of two of the four original lateral moraines, they are found where the new, combined glacier encounters the valley walls. The resulting lateral moraines of the combined glacier are therefore created from the remaining two of four original lateral moraines that did not combine to produce the medial moraine.

Terminal moraines occur at the farthermost advance of the ice. They are located near the downstream edge of the melting ice, where the rate of melting is in equilibrium with the rate of flow of ice from the source areas. At this point, rock debris is released from the ice as the ice melts, leaving a deposit of till. Such moraines can be hundreds of feet in height. It has been observed that in the case of a valley glacier, end moraines have a crescent shape that points downstream.

Continental glaciers, on the other hand, tend to leave terminal moraines that are more irregular and can extend for miles across the landscape. As the glacier recedes, it can form **recessional moraines** at points where the ice front pauses long enough in an equilibrium state to form a moraine.

Finally, **ground moraine** consists of widespread deposits of till that have been deposited over vast areas as the ice retreated. Till can exhibit a **wide range of hydraulic conductivity** values and is characterized by the fact that it is very **inhomogeneous**. In other words, the hydraulic conductivity can change markedly over very small distances. This makes modeling of till a very challenging problem.

A boulder that has been deposited on subsurface material different from the material of origin of the boulder is called an **erratic**. Erratics can be enormous, often measuring more than 10 feet in diameter. A trail of erratics leading from the source area of the boulders to their final resting place is called a **boulder train**. It can be used to determine the local direction of movement of the ice.

Stratified Deposits Deposits of materials derived from glacial ice, but water transported, are collectively called **glacial drift**. **Outwash** is generated when the meltwater from the ice encounters and entrains materials being transported by the ice and deposits them down valley from the ice margin. Streams and rivers emanating from the ice margin tend to be very fast moving and carry coarse material to be deposited as **glacial-fluvial deposits**, that is, stream-carried deposits. Outwash is characterized by coarse material interbedded and intermixed with finer-grained material such as silts and clays. A relatively permeable aquifer, outwash deposits are often used as a source of groundwater for domestic, public, and industrial uses.

Where **glacial lakes** are found, an unusual clay deposit may be formed. Characteristically, these clay deposits are made up of alternating layers of finer- and coarser-grained materials. In cross section these materials, called **varved clays**, appear striped. It is believed that these alternating layers represent summer and winter deposition. As might be expected, glacial lake clays are often found in areas near the terminus of the glacier. Some glacial lakes formed during the last Pleistocene glaciation were enormous, on the order of the size of the current North American Great Lakes.

Because lake clays have relatively low hydraulic conductivity, they are important primarily in their role as barriers to the vertical migration of contaminants in multi-aquifer systems.

At the point where a high-energy water body, such as a river, enters a quiet, low energy environment such as a lake, deposition of coarse-grained material takes place. These deposits, which have bedding that dips (or slopes) in a characteristically steep manner, are called **deltaic deposits**. Outwash and deltaic deposits associated with glacial discharge are similar in terms of grain size. However, deltaic cross-bedding can sometimes be used to distinguish one of these deposits from the other.

Several other landforms are associated with glacial activity. One of the most interesting is the **esker**. Formed by streams running through tunnels in stagnant ice, these ridges of stratified gravel (often called **stratified drift**) are created when the surrounding ice melts away. They can be from 10 to more than 50 feet in height and extend for miles. Although striking land features, they are relatively unimportant from a groundwater modeling point of view. Although they tend to form highly permeable ridges projecting above the neighboring landscape, they are normally above the water table and therefore, generally unsaturated. Thus they typically do not make useful aquifers.

Crevasse fillings are similar to eskers in many respects but tend to be less sinuous. This topography suggests that these deposits are formed by material collecting in crevasses in the ice. The deposits are left as ridges as the ice melts, similar to eskers in their mechanism of formation.

A close relative to the crevasse deposit is the **kame**, a stratified mound of irregular shape probably formed by debris collecting in openings in stagnant ice. A **kame terrace** is found between the wasting ice and the valley wall. This material is left as a terrace along the valley wall once the ice has melted.

Drumlins are landforms found largely in New England. They are ridges of an elliptic shape with the long axis of the ellipse oriented in the direction of ice flow. Because they tend to appear in clusters, localities wherein they are found are often referred to as **drumlin fields**. They are characteristically streamlined but are steeper on the upstream than on the downstream end; that is, they are steeper in the direction from which the ice flowed. Typically, drumlins are 25 to 200 ft in height.

When an ice block becomes isolated during the retreat of a glacier, it can sometimes be surrounded by till, or even buried. As the ice melts, it leaves a hole, or depression, in the till-dominated landscape. This depression is called a **kettle**. It sometimes becomes a lake or swamp and fills with organic matter. Kettles can be miles in diameter and are often recognized by the rich organic content of the soil occupying them.

1.4.2 Consolidated Rocks

When sedimentary deposits are buried for long periods of time, the materials become cemented together to form consolidated rocks. At this point the deposits lose much of their pore space and the resulting porosity can be attributed, at least in part, to fractures and dissolution. The consolidated rocks derived from the unconsolidated deposits described above are summarized in Figure 1.17.

In general, consolidated rocks are less productive as aquifers than are unconsolidated deposits, primarily because they have lower permeability. On the other hand, while the **phase average velocities**, **q**, (sometimes called *Darcy velocities*), of the groundwater may be lower in such rocks, they often exhibit low **porosity**. Low porosity may result in relatively high **pore velocity**, **v** (the velocity of fluid particles). The relationship between phase-average and pore velocities is given by

$$\mathbf{v} = \frac{\mathbf{q}}{\theta} \tag{1.4}$$

where **v** is the **pore velocity**, **q** is the **phase average velocity**, and θ is the **porosity** (more specifically, the **effective porosity**). Because flow is frequently due to **secondary permeability**, which is permeability due to such features as fractures, faults, or solution cavities, there is often significant **anisotropy**, that is, a preferential flow direction. For example, one would intuitively expect to find a larger value of hydraulic conductivity in the direction parallel to fracture orientation than orthogonal to it.

The potential importance of secondary permeability gives rise to the concept of a **double-porosity model**. Such a model has two sets of permeabilities and two sets of porosities, one associated with the secondary permeability features, such as the fractures, and the other associated with the **primary permeability** features, those of

SEDIMENTARY ROCKS				
ORIGIN		TEXTURE	PAR TICLESIZE COMPOSITION	ROCKNAME
	DETRITAL	CLASTIC	GRANULAR	CONGLOMERATE
			SAND	SANDSTONE
			SILT AND CLAY	MUDSTONE OR SHALE
CHEMICAL	INORGANIC	CLASTIC AND NON-CLASTIC	CALCITE	LIMESTONE
			DOLOMITE	DOLOSTONE
			HALITE	SALT
			GYPSUM	GYPSUM
	BIO-CHEMICAL		CALCITE	LIMESTONE
			PLANT REMAINS	COAL

FIGURE 1.17. *Sedimentary rock chart (after Leet and Judson [15]).*

the unfractured host rock. The host rock in this instance can conveniently be thought of as made up of blocks of relatively low permeability material surrounded by highly permeable fractures. Each of the two systems is described by a separate flow equation with its own set of parameters, boundary conditions, and stresses. The two systems are coupled one to the other by a leakage term that describes fluid (and contaminant mass) movement between them.

1.4.3 Metamorphic Rocks

When consolidated sediments are further subjected to intense pressure and heat, generally due to phenomena associated with mountain building, the minerals forming the rock recrystallize to form a much denser rock made up of new minerals in equilibrium with the intense temperatures and pressures under which they are now formed. The resulting rocks are called **metamorphic rocks**. While the molecular composition of the metamorphic rock is normally very similar to the original parent consolidated rock, the minerals that form the metamorphic rock are quite different. Some typical metamorphic rocks and the host rock from which they were formed are presented in Figure 1.18. Along the horizontal axis are listed the various minerals that characterize the metamorphic rocks. For example, in the second column are listed rocks identified with the chlorite metamorphic zone. The chlorite metamorphic zone rocks would normally contain chlorite minerals.

<table>
<tr><th rowspan="3">ORIGINAL ROCK</th><th colspan="5">METAMORPHIC ZONE</th></tr>
<tr><th>CHLORITE</th><th>BIOTITE</th><th>ALMANDITE</th><th>STAUROLITE</th><th>SILLIMANITE</th></tr>
<tr><th colspan="5">METAMORPHIC ROCK</th></tr>
<tr><td>SHALE</td><td>SLATE</td><td>BIOTITE PHYLLITE</td><td>BIOTITE-GARNET PHYLLITE</td><td>BIOTITE-GARNET STAUROLITE SCHIST</td><td>SILLIMANITE SCHIST OR GNEISS</td></tr>
<tr><td>CLAYEY SANDSTONE</td><td>CLAYEY SANDSTONE</td><td>QUARTZ-MICA SCHIST</td><td colspan="3">QUARTZ-MICA-GARNET-SCHIST</td></tr>
<tr><td>QUARTZ SANDSTONE</td><td colspan="5">QUARTZITE</td></tr>
<tr><td>LIMESTONE DOLOMITE</td><td>LIMESTONE DOLOMITE</td><td colspan="4">MARBLE</td></tr>
<tr><td>BASALT</td><td colspan="2">CHLORITE-EPIDOTE-ALBITE-SCHIST</td><td colspan="2">ALBITE EPIDOTE AMPHIBOLITE</td><td>AMPHIBOLITE</td></tr>
<tr><td>GRANITE</td><td>GRANITE</td><td colspan="4">GRANITE GNEISS</td></tr>
<tr><td>RHYOLITE</td><td>RHYOLITE</td><td colspan="4">FINE-GRAINED BIOTITE GNEISS</td></tr>
</table>

FIGURE 1.18. *Chart of metamorphic rocks (after Leet and Judson [15])*

Because metamorphic rocks are exceedingly dense, they have virtually no primary porosity or permeability. Thus secondary permeability due to fractures and faults constitutes the principal mechanism for groundwater-flow in metamorphic rocks. Porosity and permeability are, therefore, very low in such rock units. However, as in the case of consolidated rocks, groundwater pore velocities can be quite high, due to the relatively low porosity, as is evident from an examination of equation 1.4.

1.4.4 Igneous Rocks

Igneous rocks are formed as a result of the complete melting of existing rocks, often due to mountain building or upward convection of liquid **magma** from the earth's mantle. There are two basic types of igneous rocks, extrusive and intrusive (Figure 1.19). **Extrusive rocks** are formed when liquid magma reaches the earth's surface and forms **lava**. Lava has two basic forms, ash and flow.

Ash is the material that is extruded during volcanic activity and is carried via the wind to form ash cones. **Flow** is a liquid form of the extruded material that flows overland to form large volcanic beds.

Flows can be either very dense or composed largely of coarse, porous beds known as **aa** (**ah ah**). The dense beds may be quite porous in a formal sense, but the porosity may be irrelevant from the point of view of groundwater flow. The reason for this

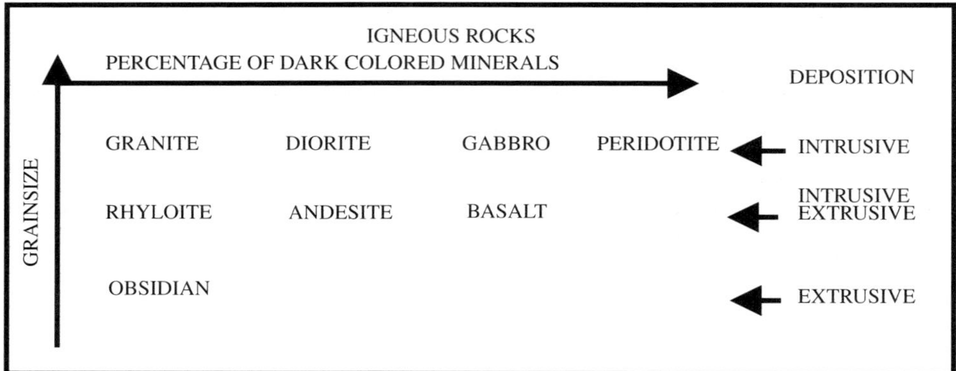

FIGURE 1.19. *Igneous rocks defined by mineralogy and grain size.*

inconsistency is the observation that the porosity is composed largely of voids generated by isolated air bubbles. Thus the pore spaces are unconnected and therefore ineffective for the transmission of water.

Enormous areas of the northwestern United States are covered by **basaltic flows**. Often, the basaltic flows consist of interbedded layers of dense and porous basalt. The porous basalt layers generally form good aquifers that yield considerable volumes of water to wells. The dense basalt is capable of limited groundwater flow, due primarily to secondary permeability.

Intrusive rocks are generally called **igneous rocks**, although this is formally a misnomer. Intrusive rocks consist of the very dense rocks that crystallize below the surface of the earth. Those that form at great depths are most often **granites**, although other coarse crystalline rocks classified on the basis of their grain size or mineral composition are found widely. In general, granitic rocks have virtually no primary porosity or permeability. Secondary permeability and porosity are due to fracturing and faulting.

Intrusive rocks normally appear on the earth's surface through the **erosion** of overlying rock units. For example, the rocks that form the core of mountain chains, that is, the rocks that are found at great depth beneath mountains, can, through erosion of the overlying rock units, make their way to the earth's surface.

However, intrusive rocks can also appear on the earth's surface through injection and subsequent crystallization of liquid magma along fractures and faults. Molten rock will travel along fractures, whether horizontal or vertical. As the molten rock approaches the earth's surface, it crystallizes to form dense rock masses, generally referred to as **basalt intrusive rocks**. When the fractures occupied by these rocks are vertical, the rock formations are called **dikes**. When the intruded fractures are horizontal, the intrusions are called **sills**. In either case, the contact between the basalt and the host rock can be very permeable, due to the existence of a zone of fractures due to cooling that occurs between the molten and host rock. Such fractured zones can provide important secondary permeability. Contaminant can travel significant distances along these high hydraulic–conductivity zones.

1.4.5 Representation of Geological Units

It should be evident at this point that general knowledge of the geology of an area to be modeled is important. To understand the geology often requires that one organize available information into forms usable by the groundwater professional. Two such forms often encountered are the **cross section** and the **interfacial contact contour map**.

Geohydrological Cross Section The **cross section** and its close cousin the **fence diagram** are generated from subsurface information. The most common form of subsurface information is the **well or boring log**. A hypothetical example is found in Figure 1.20. The boring log records both quantitative and qualitative information gleaned from information obtained at the well site during the drilling operation. Information regarding the groundwater company supervising the boring operation and other information relevant to the log of this boring is indicated at the top of the diagram and in the lower right-hand corner. Examining the columns left to right, we observe the following: The first column indicates the depth in feet below the land surface, the second indicates where soil samples were taken for analysis, and the third records the depth to the top and bottom of the soil samples.

The next two columns record measured concentrations of contaminants. The top number of each pair represents the total purgeable hydrocarbon concentrations, and the lower one describes the total extractable hydrocarbon concentrations. The purgeable compounds are volatile compared to the extractable compounds. The fifth column is labeled "TLC results." TLC is the acronym for *thin-layer chromatography*, a method of chemical analysis that can be used in the field to get a rough estimate of the contaminant concentration in the sample. It is very helpful in determining the relative level of concentrations between samples. Thus it can be used as a guide in determining which samples should be sent to the laboratory.

The sixth column is a visual representation of the material observed by the groundwater professional and described in columns 7, 8 and 9 indicating the construction of the well, and finally, the last column records the elevation of points in the boring (in contrast to the depth to points shown in column 1).

To generate the cross section, a line is drawn on a base map that generally passes close to the wells to be used for the cross section. A hypothetical example is given in Figure 1.21.

Line A–B passes through some well locations and close to others. To generate a cross section, the geohydrological information obtained from these wells is projected to line A–B and presented as a cross section. One representation of a cross section indicative of information found along line A–B identified in Figure 1.21 is given in Figure 1.22. The two geological units encountered in the logs for the wells indicated can be identified by the two patterns appearing on this cross section.

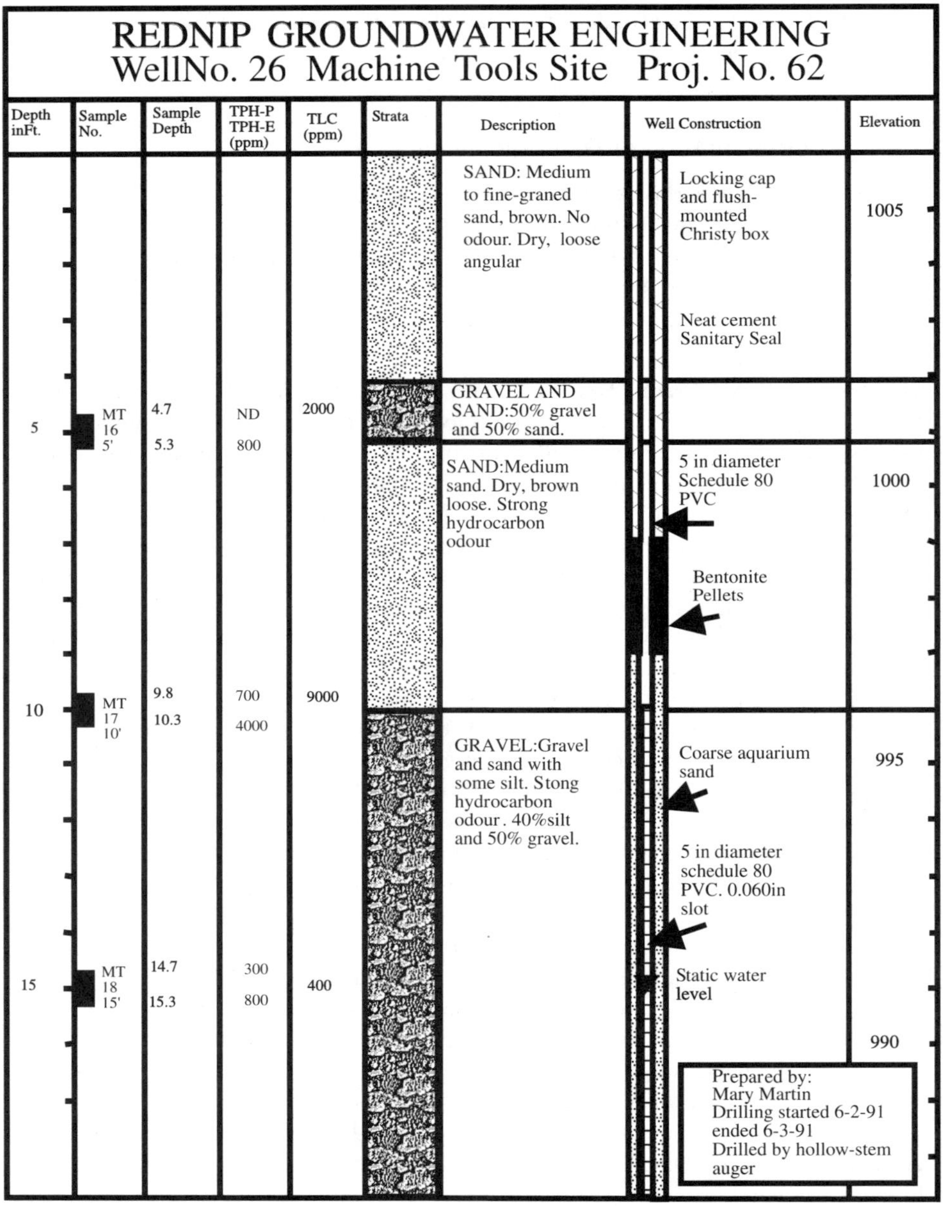

REDNIP GROUNDWATER ENGINEERING
WellNo. 26 Machine Tools Site Proj. No. 62

Depth inFt.	Sample No.	Sample Depth	TPH-P TPH-E (ppm)	TLC (ppm)	Strata	Description	Well Construction	Elevation
						SAND: Medium to fine-graned sand, brown. No odour. Dry, loose angular	Locking cap and flush-mounted Christy box Neat cement Sanitary Seal	1005
5	MT 16 5'	4.7 5.3	ND 800	2000		GRAVEL AND SAND:50% gravel and 50% sand.		
						SAND:Medium sand. Dry, brown loose. Strong hydrocarbon odour	5 in diameter Schedule 80 PVC Bentonite Pellets	1000
10	MT 17 10'	9.8 10.3	700 4000	9000				
						GRAVEL:Gravel and sand with some silt. Stong hydrocarbon odour. 40%silt and 50% gravel.	Coarse aquarium sand 5 in diameter schedule 80 PVC. 0.060in slot	995
15	MT 18 15'	14.7 15.3	300 800	400			Static water level	
								990

Prepared by:
Mary Martin
Drilling started 6-2-91
ended 6-3-91
Drilled by hollow-stem auger

FIGURE 1.20. *Hypothetical boring log.*

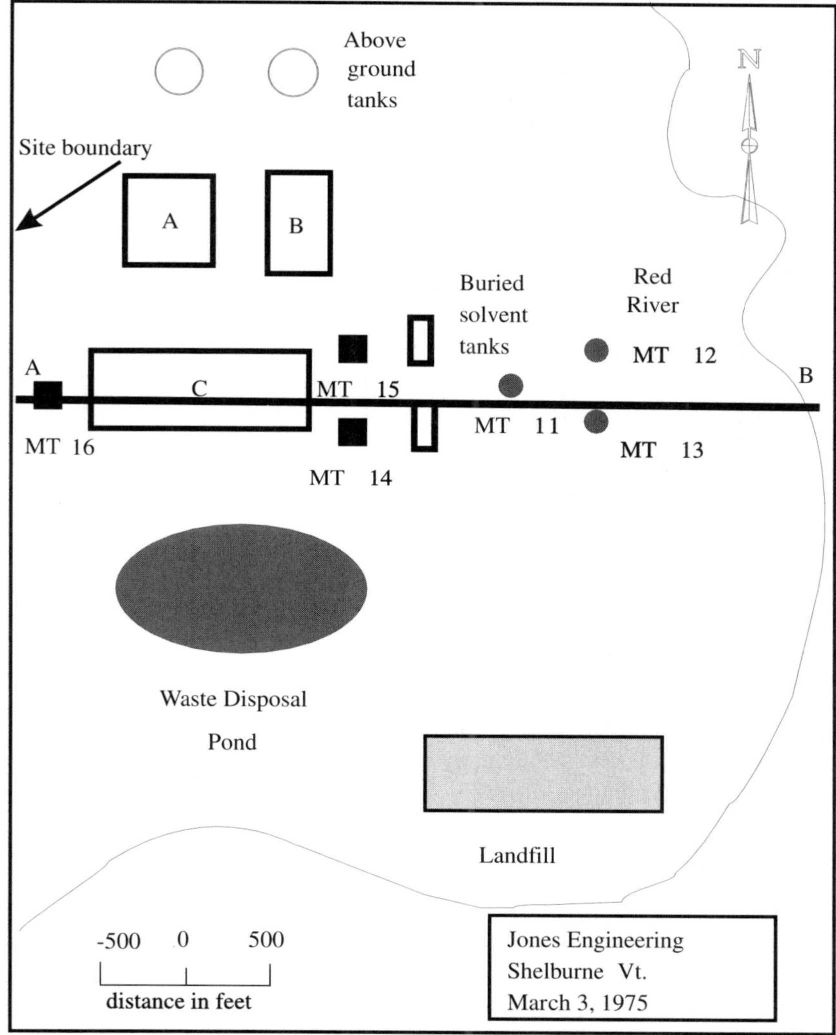

FIGURE 1.21. *Location of cross-section.*

The **fence diagram** is a three-dimensional variant on the cross section and is made up of several cross sections connected together in a pattern resembling farmers' fences as seen from an airplane—thus the name *fence diagram.*

While the analysis above gives insight into the **stratigraphy** of a geological sequence and, to a certain degree, the extent of various geological horizons, this is not the specific information required by a groundwater-simulation program. The model needs to know the elevation of the interfaces that separate various geological hori-

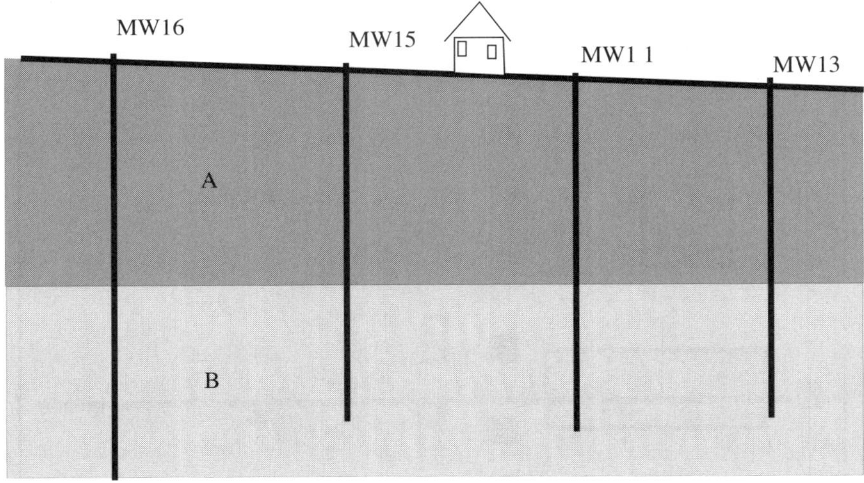

FIGURE 1.22. Typical cross section as identified in the preceding map.

zons. Thus this information must be established via the well logs, subsequently contoured, and then introduced as input into the groundwater-flow computer program.

Various algorithms and their concomitant programs are available to assist in providing this information. **Kriging**, which is, in essence, an interpolation program that accommodates uncertainty and can provide estimates of the variance of the parameters being interpolated, is commonly used for the generation of **interfacial surfaces** (and therefore contours).

Tucson Example

Let us now consider the discussion above within the context of the Tucson example. In preparation for the formulation and implementation of a groundwater-flow and transport model of the area of interest, the Arizona Department of Water Resources organized, interpreted and quantified the available geological information (see Mock et al. [6]). They describe their approach as follows:

> The data used for the geologic analysis was in various forms and from multiple sources. Data, in the form of drillers' logs, drill cuttings analyses, geologists' logs and geophysical logs, were collected from the United States Geological Survey (USGS), Arizona Department of Health Services (ADHS), Arizona Department of Water Resources (ADWR), Tucson Water (TW), Hargis and Montgomery, Inc. (H&M), Black and Veatch, Inc. (B&V), and Ecology and Environment, Inc. (E&E).[18] This information was used to prepare geologic cross sections for the interpretation of the hydrogeology.

[18]Hargis and Montgomery, Black and Veatch, and Ecology and Environment are private consulting firms that have worked in various capacities at the site over a number of years.

> A classification system for sediment and rock type was devised to integrate the information provided from the drillers' and geologists' logs and drill cuttings analysis. Five categories were created which separate clay (fines), from sandy-clay, from clayey-sand, from sands and gravels, from hard rock.... The clay (fines) category encompasses silt, and further references to clay in the main text and this appendix, includes silt in the classification of clay (fines). The categories were color coded. Sandy-clay and clayey-sand were considered transitional sediment categories, representing the gradation from clay to sand and gravel. Lithologic columns were then constructed utilizing the classification system for the selected wells.

Let us now consider the nature of the logs mentioned in this paragraph. We begin by quoting from Schmidt [9], wherein he describes the nature of the boring and sampling protocol.

> Twenty-one groundwater monitor wells were constructed under contract to Tucson Water... between April 26, 1984 and August 31, 1984 with funds provided by the Arizona Department of Health Services (ADHS) under Environmental Protection Agency (EPA) guidelines for Superfund project investigations. These wells were located and constructed to test for the presence of trichloroethylene (TCE) contamination of the groundwater....
>
> Western Well and Pump of Colby, Kansas was the successful bidder on specifications developed jointly by the Remedial Investigation Team (RIT). Reverse circulation rotary drilling was required to a maximum depth of four hundred feet, using only water for makeup of the drilling fluid. The project specified the placement of monitor wells within public right-of-ways to eliminate the need to purchase land for well sites. This required the driller to use above ground tanks or pits to eliminate digging around existing utilities in the right-of-ways. The specifications also required the contractor to drill dry to the water table in three locations, using only air to remove the cuttings. Samples of the formations encountered during each five foot interval of drilling, by both methods, were collected by Tucson Water personnel. Grain size analyses were performed and lithology described on all samples of drill cuttings that were collected.
>
> The U.S. Geological Survey (USGS) geophysically logged several of the boreholes under an Intergovernmental Agreement (IGA) with Tucson Water and ADHS. Reduced scale drawings of the geophysical logs were provided for the boreholes that were logged. The USGS also constructed grain size histograms from the well cuttings analyses provided by Tucson Water. All geologic drilling data, geophysical logs and well construction information were provided to the RIT as soon as processing was completed.
>
> As mentioned above, three sites were selected for drilling dry with air to the water table. The purpose of this drilling method was to determine the presence or absence of perched water above the regional water table. Dry drilling began on April 26, 1984 using a tricone bit and dual tube drill pipe. Cuttings were returned via the drillpipe annulus and routed through a cyclone catcher to dissipate their energy. Cuttings were collected over each five foot interval and logged by a Tucson Water representative on the site. Grain size analyses were performed on each five foot interval and a lithologic description made of the washed sample. All three of the dry boreholes were later drilled

deeper with the reverse circulation rotary drill rig using City water for makeup of the drilling fluid for removing the cuttings.

The construction of all wells began with the augering of a twenty-four inch diameter hole to a depth of twenty feet. Sixteen-inch surface casing was then placed in the augered holes and cemented in place. The reverse circulation rotary rig was then positioned over the surface casing to drill a twelve-inch borehole to the depth specified by Tucson Water.

A representative of Tucson Water was present during all drilling operations to collect and describe the drill cuttings and to monitor drilling procedures. The cuttings were collected in a sample catcher attached to the above ground portable tanks. At the end of each day of drilling, or the termination of the borehole, samples were taken to the Tucson Water soil laboratory for processing. The individual samples were wet washed into three size fractions, 0.062mm, 2mm, and 74mm. The weight percentage of each fraction was then calculated.

Selected boreholes were geophysically logged by the USGS immediately after the drill stem was removed. Geophysical logs run were caliper, neutron, natural gamma, gamma-gamma, short and long normal resistivity. The device used was a Mt. Sopris portable unit mounted in a 3/4 ton van. Logging generally required four to five hours to complete.

As soon as the logging was completed and interpreted, or if no logging was performed on a particular borehole, as soon as the drill stem was out of the borehole, a casing/screen diagram was developed by the Tucson Water hydrologist and given to the driller. The screen and casing were then assembled according to this diagram and lowered into the borehole. A gravel pack was placed around the screened section. Gravel was pumped through a one and one-half inch tremie pipe to the specified depth. A 5-foot layer of fine sand was then tremied in on top of the gravel pack. This sand layer formed a seal to prevent the cement grout from penetrating into the gravel pack. Cement grout was then pumped, via the tremie pipe, until the borehole around the blank casing was filled to within two feet of the land surface.

Six-inch diameter schedule 80 PVC casing was used, with butt-threaded steel couplings to connect the casing joints. The screened sections had 0.040 inch slot openings, spaced 0.125 inches apart. There were 72 slots per row and 8 rows per foot of casing. This gave an open area of approximately 20 per cent. Centralizers were placed every 60 feet, beginning at the bottom, to keep the casing vertical and centered within the borehole. This allowed placement of an effective grout seal for each monitor well.

The criteria for selection of the screened interval for shallow wells were the locations of zones of coarse sands and gravels below the water table. Screen sections were placed opposite these coarse sand and gravel zones. Every effort was made to screen from no more than 10 feet beneath the water table to the top of a clay unit, which was identified throughout the study area. This produced shallow wells ranging from one hundred thirty to two hundred forty five feet in depth and screened over thirty to one hundred foot depth intervals. Three deep wells were screened from three hundred feet to four hundred feet to test for TCE in the groundwater beneath the clay unit. These deep wells were constructed adjacent to shallow wells.

Twenty four hours or more after construction was completed, the wells were developed by air lifting and surging. A conventional rotary drill rig was used for well development. Three-inch drillpipe was assembled and placed inside the cased well. Compressed air was then circulated down the drillpipe forcing water up the annular space. The compressed air was shut off at intervals when the water became clear, allowing the water level to recover. The air was then turned back on to produce more water. This procedure continued until the water remained clear in between surgings. During this stage of development the wells produced from thirty to one hundred gallons per minute.

A temporary test pump was installed in each of the wells after development. The wells were then pumped for twelve hours each while being monitored and sampled by personnel from the ADHS and Tucson Water. Static water levels were measured before the start of each test and pumping water level readings were taken during each test. Water level recovery was then measured after pumping had stopped. Plots of drawdown and recovery readings versus time were prepared by ADHS.

Permanent underground concrete vaults were constructed around each of the well heads to protect the well and controls from vehicle traffic and vandalism. The vaults utilized were traffic grade utility splice boxes with metal lids designed to withstand light vehicular traffic. The vaults were set over the wells with their metal lid just above the existing grade.

Permanent pumps designed to produce forty to sixty gallons per minute were installed in each well. The shallow wells were equipped with three horsepower motors, while the deep wells were equipped with five horsepower motors. Final acceptance tests lasting between four and six hours each were conducted on all wells. Water quality samples and water level measurements were also taken during these tests. Upon final completion and acceptance, these monitor wells were included in the sampling network of the TAA study program.

Use of Boring Information Several methods are used to communicate the information obtained from a boring. A **written log** is prepared in the field at the time of drilling. An example is provided in Figure 1.23. The information provided is similar to our hypothetical log provided as Figure 1.20. The information at the top of the table identifies the type of log being recorded and the boring identification. The first and second columns record the depth and depth interval, respectively, that is to be associated with the information that appears in the remaining two columns in this table. Column three describes the material observed by the groundwater professional, and the fourth column provides a more detailed description when appropriate.

In Figure 1.24 information derived from observations made in the field and in the laboratory are presented graphically. The drawing on the left represents the grain-size analyses conducted on the samples collected on site. The remaining graphs represent the results of various logs that were made at the time the boring was completed. The first of these is called a **caliper log**. It records the diameter of the boring as a function of depth. The variation in diameter can provide insight into the **lithology**[19] of the subsurface since some materials are more stable than others during the drilling process. For example, coarser material often is less stable and therefore falls into the

[19] *Lithology* refers to the physical characteristics of geological formations, such as grain size and color.

Table B-19

Lithologic log for SF-19 Monitor Well.

DEPTH INTERVAL (feet)	INTERVAL THICKNESS (feet)	DESCRIPTION OF GEOLOGIC MATERIAL GENERAL	DETAILED
25-45	20	Gravelly sand with caliche:	Gravelly sand with caliche
45-55	10	Clayey sand:	Clayey sand
55-60	5	Sandy gravel with cobbles:	Red, sandy gravel, sedimentary and volcanic
60-65	5		Red, sandy gravel, sedimentary and volcanic
65-70	5		Dark brown, sandy gravel, cobbles
70-75	5		Dark brown, sandy gravel
75-80	5		Cobbles
80-85	5		Very sandy gravel
85-90	5		Very sandy gravel
90-95	5	Clay, sandy clay and silty clay:	Light brown fine sandy clay
95-97	2		Dense, sticky clay
97-105	8		Medium-fine sand
105-115	10		Dense, silty clay
115-119	4		Dense, sticky clay
119-136	17	Clayey sand and gravel:	Coarse gravel and sand
136-198	62	Clayey sandy gravel:	Sandy, gravelly clay
198-230	32		Brown, gravelly sand
230-263	33		Brown, gravelly sand
263-322	59	Clayey sandy gravel:	Brown, clayey, sandy gravel
322-405	77	Clayey gravelly sand:	Clayey sand and gravel

Total depth of borehole = 405 feet.
Logged by Tucson Water, City of Tucson

FIGURE 1.23. *A typical written log derived from information obtained from a boring log at the Tucson site [6].*

boring, resulting in a larger-diameter hole at the depth at which the coarse material resides.

The remaining four graphs record the results of **geophysical logs**. Two are associated with the **resistivity** of the various formations encountered with depth. High resistivities are indicative of dense rocks, such as granite and some limestone, while medium to high values may indicate saturated sands. The curves presented in Figure 1.24 indicate two resistivity curves. One records conditions very near the borehole, and these measurements may be influenced by drilling activities. The second provides information on materials more distant from the borehole.

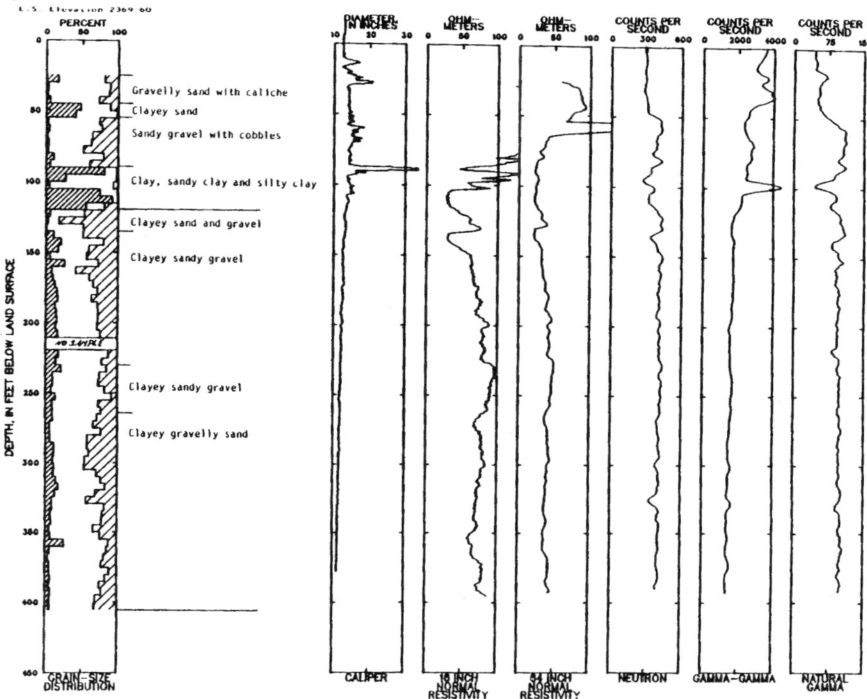

FIGURE 1.24. Diagrammatic representation of information derived from borehole observations and laboratory analysis.

The fourth curve is a **neutron log**. It is a measure of the slow neutrons induced by the geological formation in response to bombardment by high-energy neutrons emitted from an artificial radiation source introduced into the boring by the groundwater professional. Hydrogen has the greatest impact on the response of the formation to neutron radiation. Thus the principal use of the neutron log is to identify those formations that contain the greatest proportion of water. Given that the formations are saturated, this is also a measure of the formation porosity.

The last two curves are gamma logs. The **natural gamma log** records the natural radioactivity of the various materials penetrated by the boring. Because different formations will contain different proportions of radioactive compounds, especially uranium or thorium, the gamma log can be used to differentiate between lithologic layers. Natural gamma radiation is normally high in clays and shales because of their mineral composition. Sands and sandstones, on the other hand, tend to produce less natural radiation.

The **gamma-gamma log** uses an artificial gamma source similar to the neutron log and measures the gamma radiation that reaches a detector that is shielded from the source. The amount of radiation detected depends on the density of the formation. Thus, at least in concept, given a knowledge of the grain density, the bulk density, and the fluid density, the fluid content as well as the porosity can be established.

Use of Cross Sections The locations of 12 preliminary geological cross sections were selected and delineated on a base map. Five north–south and seven east–west cross sections were constructed using the color-coded lithologic columns for the wells selected. As described by Black and Veatch [10]:

> A **lithofacies map** was constructed for the TAA from the preliminary geologic cross sections. This map showed the approximate extent (depth-related) of the aquitard, the deep fine-grained facies and the transition from a clay to a sandy-clay within the deep fine-grained facies, and the extent of the coarse-grained facies.

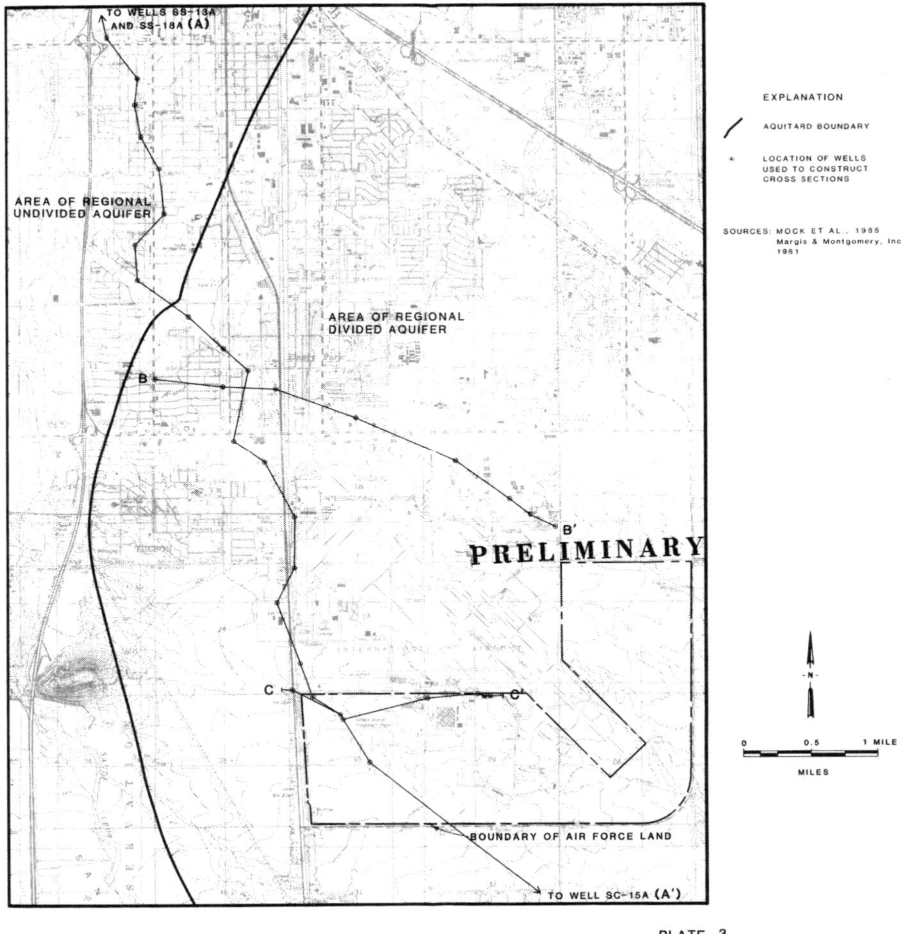

FIGURE 1.25. *Location map for cross sections. The section of interest is represented by the line with the endpoints A and A′. Modified from Black and Veatch [10].*

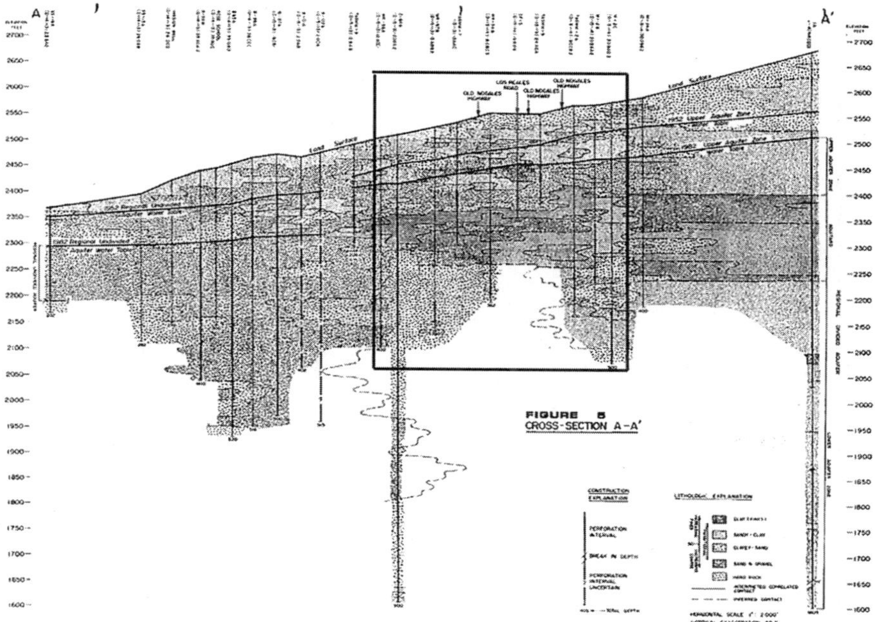

FIGURE 1.26. *Cross section along the profile A-A′ illustrated in the definition figure. The shaded area represents approximately the aquitard. The boxed area is expanded to a larger scale in Figure 1.27. Modified from Mock et al. [6].*

> Two **geologic cross-sections**, one north–south and one east–west, were chosen from the 12 preliminary cross sections (see Figures 1.25 to 1.27). These cross-sections were reanalyzed to refine the sedimentary trends originally established with the preliminary cross-sections. Well records were reviewed again to obtain additional logs in the proximity of the cross-sections which would provide greater geologic detail for correlative purposes. The location of the final cross-sections were slightly altered to incorporate the superior lithologic data which was supplied by the WR and SF monitor well series. The additional detailed geologic cross-sections... were constructed utilizing the SF data. When available, geologic logs used in conjunction with the drill cuttings analysis provided excellent lithologic detail for geologic analysis.
>
> A plot of subsurface hardrock elevations on a plan-view map of the study area revealed north–south trending elevational features. **Hard rock**, as defined in the *Dictionary of Geological Terms* and in this text is "loosely used to distinguish igneous and metamorphic from sedimentary rock." A survey of well logs for depth and description other than those of sediments was conducted. The hard rock elevational levels were based on the location of the wells on the plan-view map and the elevation at which hard rock was reported.

It is appropriate at this point to revisit the information provided in Figures 1.25 to 1.27. The bold line that defines the cross section A–A′ is a piecewise linear curve

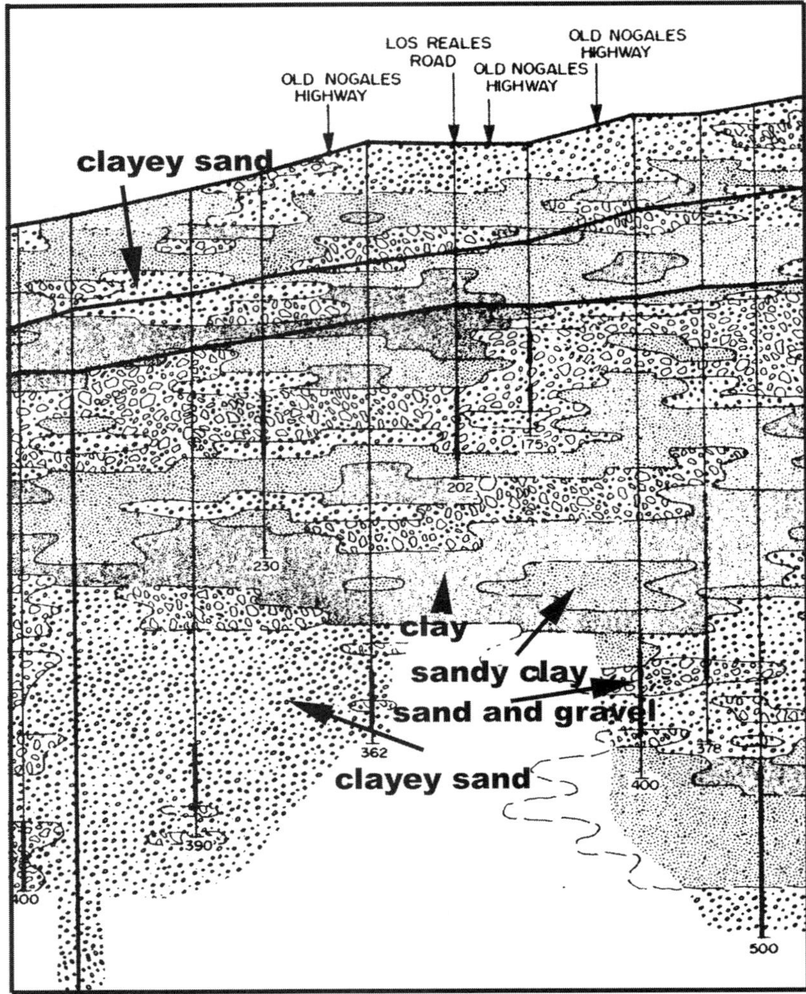

FIGURE 1.27. Section expanded to larger scale to illustrate detail of stratigraphy and to provide identificaton for patterns identified with geohydrologic properties.

that connects borings from which the information provided in Figures 1.26 and 1.27 was derived. Thus this line has been stretched taut to create the cross section. The vertical lines that extend from the surface to various depths are the borings. Each is identified by the label found above the land surface depicted, in vertical alignment with each boring. Also evident in Figure 1.27 is the identification of discernible landmarks, in this case major roads.

The cross section records the materials found in the borings, and this information is recorded as different material property-identifying patterns. Figure 1.27 illustrates this more clearly, along with information as to the depths at which the resulting wells are screened (in contact with the geological formations). Also presented in this figure are the various materials found in the cross section; these are clay, clayey sand, sand, sandy clay, and sand and gravel. The remaining pattern is that of hard rock, which is not shown in Figure 1.27.

Note the two lines approximately parallel to and 150 feet below the land surface. The upper one of these two lines represents the water table in the **upper aquifer unit** (above the aquitard) and the lower one the water table in the **lower aquifer unit** (below the aquitard). The difference in these water levels is due to greater pumping in the lower aquifer than the upper. Also of interest is the dislocation, or sudden drop, in the elevation of these two curves as one enters an area where the aquitard is missing.

Based on the information collected in the investigation above, Mock et al. [6] describe the **regional geological setting** as follows:

> The TAA is located in the west-central portion of the Tucson basin of southeastern Arizona. This basin trends northwesterly and is roughly triangular in shape. Elevations of the surrounding mountains range from 3000 feet in the west to over 9000 feet in the northeast and south. The basin floor has an elevation of approximately 2000 feet at the northwest end and rises to 2900 feet in the south.
>
> The mountain ranges which border the Tucson basin are the Santa Catalina Mountains to the northeast and the Rincon Mountains to the east. Both ranges are composed of metamorphic core complex granodiorites, part of which were formed in the mid-Tertiary Orogeny.... The Santa Rita Mountains border the basin to the south east and are composed of Triassic and late Cretaceous volcanic and sedimentary rocks. The Sierrita Mountains, to the south west, are Triassic to Cretaceous rhyodacites and monzonitic rocks. Black Mountain is located on the western edge of the basin and is composed of late Oligocene and Miocene andesites. The rhyodacites and andesites (late Cretaceous to Paleocene) of the Tucson Mountains form the northwest boundary for the basin (Drewes [8]).
>
> The rock fragments contained in the alluvial fill of the TAA, as described from drill cuttings, are composed of basalt, andesite, latite, tuff, rhyolite, granite, gneiss, and quartzite rock fragments. Volcanic rock fragments may have originated in the Tucson Mountains and/or Black Mountain to the west, whereas granitic and gneissic rock fragments may have had their source in the Santa Catalina and Rincon Mountains to the northeast and east, or from mountain ranges to the south.

Although the information above is important in conceptualizing the geology of the aquifer, the primary geological input to the model is the distribution of geological materials in three space dimensions. Information provided in the horizontal, or areal, plane provides insight into the distribution of the various **hydrogeological** and **hydrogeochemical** model parameters. In the vertical dimension this is also true, but the vertical variations in these parameters tend to be relatively sharp and distinct. Thus the vertical lithology is important.

Formation tops and bottoms constitute important parametric input. The tops define the elevations of the tops of the various formations, and the bottoms define the elevations of the bottoms. Thus the difference between the tops and bottoms provides the *formation thickness*. The tops and bottoms surfaces are generally provided as contours by the groundwater professional and must be converted to computer-readable form for input into the model.

In Argus ONE, one can input hydrogeological information in several ways. At the most primitive level, information can be entered as a constant value over the model layer (in the sense of a layer of elements). Let us consider the option of defining one global value for the elevation of layer 1.

Figure 1.28 illustrates the procedure. Selecting *Layers* from the *View* option on the menu bar creates a window that permits the selection of layers. The result is the *Layers* window shown in Figure 1.28. One then selects from the options *Bottom Elevation*. The uniform global elevation is recorded under the column denoted as *Value* using the appropriate units. To modify this value, one clicks on the f_x button, whereupon a dialog window opens that contains the current value and allows for its modification.

Although it is convenient to assign a **global value to the elevation** of the bottoms of formations, this is not normally especially realistic physically. The surfaces of geological formations typically are irregular, as is the earth's surface today. Thus a convenient way to input a variable formation topography is needed.

A variable input for formation elevation is normally achieved as follows. Make the *Bottom Elevation* layer active. From the toolset available in the upper left-hand corner of the window, select the *Contour* tool. Using this tool in a manner analogous to that used to define the domain outline, define a contour to be associated with a known elevation. When you arrive at the last point of the contour and double-click

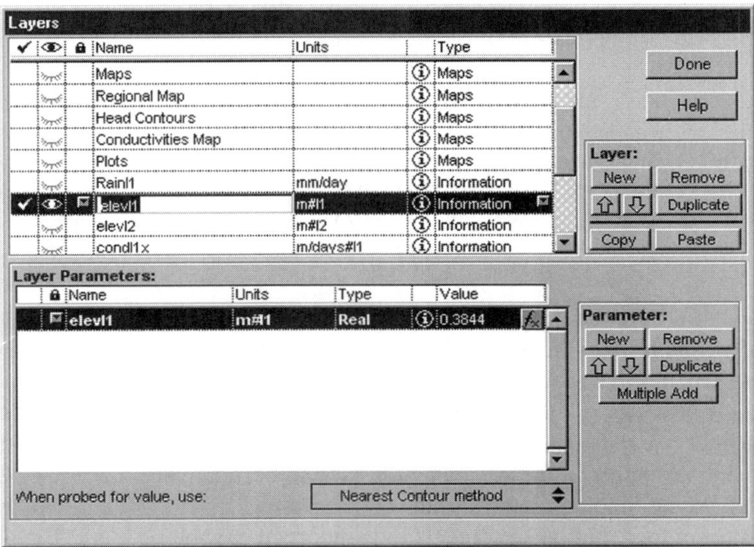

FIGURE 1.28. *Argus ONE window used to define global value of layer elevation. The highlighted value is now assigned to the entire layer.*

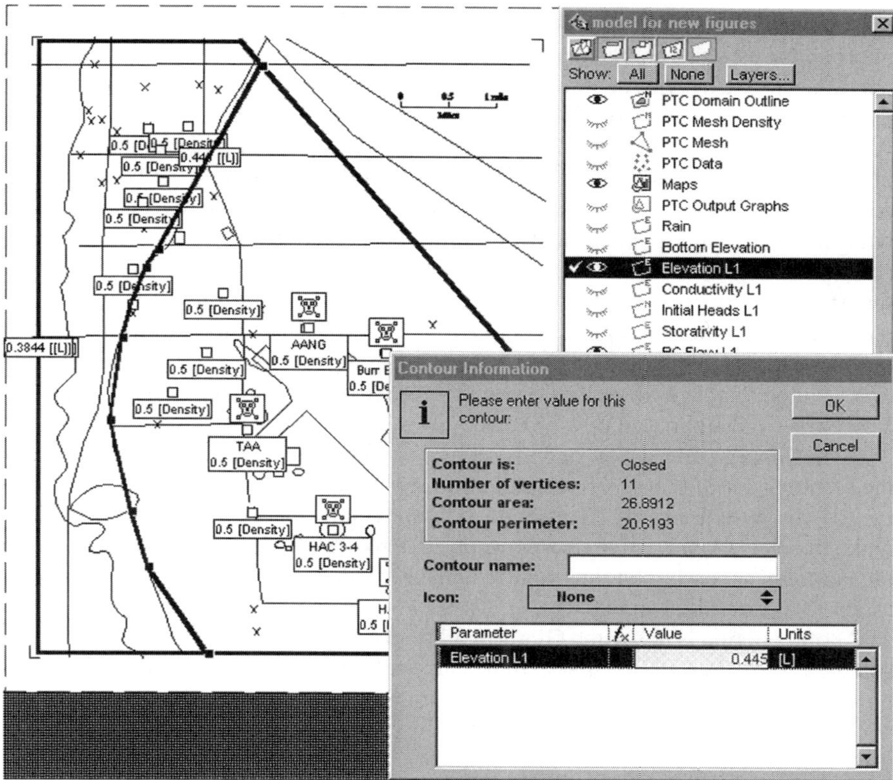

FIGURE 1.29. A contour elevation is obtained by using the contour tool. The value of the elevation is assigned via the smaller window which opens once the contour being drawn is closed via a double click on the last node.

to close it, a dialog window opens that indicates the elevation to be associated with the contour you have just defined. This number can be changed to that appropriate for the just-completed contour. The area encompassed by the contour will have the elevation denoted for the contour (see Figure 1.29).

Another option that can be helpful in defining a variable thickness involves use of the *Interpolation method* option. To utilize this feature, from the *Layers* dialog box click and hold on the spin box designated as *Nearest Contour method*. Select from the resulting menu the *Interpolation method* option. Return to the active layer window and select the *Contour tool* (either the point, open contour, or closed contour). Click on the locations where you wish to assign a specific thickness value. In the case of a contour, complete the closed or open contour. A dialog box appears allowing you to specify either a point or a contour value. When this box is closed, an interpolated thickness field incorporating the new information is created immediately. This can be verified by moving the cursor over the field of interest and observing the change in values appearing in the lower left-hand corner of the window.

Note that when the interpolation method is used with open or closed contours, the interpolator uses the points that define those contours for the interpolation. Thus if one wants to use a series of straight lines as contours, each line should be drawn with several points, not just two. Otherwise, the interpolation will be poor.

Interestingly, unlike the *Exact Contour method*, it is permissible to have overlapping contours when using the *Interpolation method*. In addition, whether or not the contours overlap, the values within a contoured region will change to be consistent with the overall interpolated field. A simple but effective way to create computer-readable contours such as described above is by using a scanned copy of contours reported in the published literature. One can import into a *Maps* layer a *.dxf* file or a *.gif*, *.tif*, or *.bmp* file of a scanned contour map of elevations. Then, using the contours on this map as a guide, trace new contours using the *Contour tool*. The traced contours can be assigned the values reported in the literature (on the scanned diagram). These traced contours can now be read by Argus ONE.[20]

To accomplish the above, one first activates the *Map layer*. Select *File*, and from the resulting set of options, select *Place Image*. From the ensuing dialog box, select the file you wish to import. The imported figure will now appear on the screen. Next, make the *elevl1* layer active. It is now possible to use the *Contour* tool to trace over the contours appearing on the imported picture. Values for the contours can now be assigned as described above.

Another option is to use the GIS capabilities of Argus ONE to input the contoured surface information from suitable third-party software. We will see in later sections how the information input at contours is communicated to the model.

1.5 COMPILATION OF HYDROLOGICAL INFORMATION

Groundwater models require specific **geohydrological information**. As mentioned earlier, one must specify geohydrological parameters, boundary conditions, and hydrological stresses. The specific requirements can best be understood in terms of the fundamental groundwater flow equation, that is,

$$\nabla \cdot \mathbf{q} = -S_s \frac{\partial h}{\partial t} - Q \tag{1.5}$$

$$\mathbf{q} = -\mathbf{K} \cdot \nabla h \tag{1.6}$$

or

$$\nabla \cdot \mathbf{K} \cdot \nabla h = S_s \frac{\partial h}{\partial t} + Q \tag{1.7}$$

where $\mathbf{K}$ is the **hydraulic conductivity**, h is the **groundwater or hydraulic head** (often identified synonymously with *water-table elevation*), S_s is the **specific storage**, and Q is a **sink term**, such as a well. By changing the sign of Q, a source term

[20]One can create as many project layers as needed by clicking *New* under *Layer* and assigning different types of layers, that is *Maps*, etc.

is generated such as could be identified with net infiltration from precipitation. The hydraulic conductivity is in bold because it is a tensorial property. That is, in general, it exhibits directional properties. It has nine coefficients in a two-dimensional system.

The hydraulic conductivity can therefore be represented by

$$\begin{bmatrix} K_{xx} & K_{xy} & K_{xz} \\ K_{yx} & K_{yy} & K_{yz} \\ K_{zx} & K_{zy} & K_{zz} \end{bmatrix}$$

When any one of the three diagonal components of this tensor are different from the others, the aquifer is called **anisotropic**. By the same token, when these components are the same, the aquifer is called **isotropic**.

When all of the components of the hydraulic conductivity matrix are constant in space, that is, the values do not vary from point to point over the aquifer, the aquifer is called **homogeneous**. When the values are a function of space, the aquifer is called **nonhomogeneous** or **heterogeneous**.

To see the impact of this representation on the system, let us write the full form of Darcy's law as

$$\begin{bmatrix} q_x \\ q_y \\ q_z \end{bmatrix} = -\begin{bmatrix} K_{xx} & K_{xy} & K_{xz} \\ K_{yx} & K_{yy} & K_{yz} \\ K_{zx} & K_{zy} & K_{zz} \end{bmatrix} \cdot \begin{bmatrix} \dfrac{\partial h}{\partial x} \\ \dfrac{\partial h}{\partial y} \\ \dfrac{\partial h}{\partial z} \end{bmatrix}$$

The component of flow in the x direction can now be shown through matrix-vector multiplication to be

$$q_x = -\left(K_{xx}\frac{\partial h}{\partial x} + K_{xy}\frac{\partial h}{\partial y} + K_{xz}\frac{\partial h}{\partial z} \right) \tag{1.8}$$

Equation 1.8 shows that the flow in the x direction is dependent in general on all three components of the head gradient when an anisotropic hydraulic conductivity is employed. The occasion when one most often employs an anisotropic form of the hydraulic conductivity is in a cross-section model when significant layering of geological formations is encountered. In such situations the horizontal components of hydraulic conductivity are normally greater that the vertical, often by more than a factor of 10.

Let us now consider in turn each of the geohydrological parameters required by the model.

1.5.1 Geohydrological Parameters

In a groundwater flow model, the following parameters are important:

1. *Hydraulic conductivity* $\mathbf{K}$ (often used synonymously, and incorrectly, with *permeability*)
2. *Specific-storage coefficient* S_s (a measure of the elasticity of a porous-medium system)
3. *Thickness of the stratigraphic layers* in an areal two-dimensional model

These parameters must be specified everywhere in the groundwater system. In practice, this means that a value must be specified for each nodal location in the model (the specification of nodal locations is addressed later). In the case of a two-dimensional areal model, the hydraulic conductivity and aquifer thickness can be replaced by the single parameter transmissivity $\mathbf{T}$, where

$$\mathbf{T}(\mathbf{x}) = l(\mathbf{x})\mathbf{K}(\mathbf{x}), \qquad \mathbf{x} \in \mathbf{\Omega},$$

$\mathbf{\Omega}$ is the areal representation of the aquifer formation, and $l(\mathbf{x})$ is the aquifer thickness.

1.5.2 Boundary Conditions

Boundary conditions must be specified at all points along the boundary, or perimeter, of a model, as defined in the Tucson example in Section 1.2. In general, the perimeter will be a surface because a three-dimensional model requires that boundary conditions be specified along the top and bottom of the model as well as along the sides. If the model is two-dimensional, boundary conditions must be specified only along a line defining the perimeter provided in Section 1.2. If there is more than one defining surface, as would be the case, for example, if the aquifer were doughnut-shaped, boundary conditions would be specified along more than one surface.

Boundary conditions can have three forms: **Dirichlet or constant head**, **Neumann or constant flux**, and **Robbins or induced flux**. Mathematically, the boundary conditions are stated as

$$h(\mathbf{x}) = h_0(\mathbf{x}), \quad \mathbf{x} \in \partial\mathbf{\Omega}_1 \qquad \textit{Dirichlet} \tag{1.9}$$

where h_0 is the specified head along the boundary segment $\partial\mathbf{\Omega}_1$ of the modeled domain $\mathbf{\Omega}$.

$$\frac{\partial h(\mathbf{x})}{\partial \mathbf{n}} = \frac{\partial h(\mathbf{x})}{\partial \mathbf{n}}\bigg|_0, \quad \mathbf{x} \in \partial\mathbf{\Omega}_2 \qquad \textit{Neumann} \tag{1.10}$$

where $\partial h(\mathbf{x})/\partial \mathbf{n}|_0$ is the specified outward normal gradient to the boundary segment $\partial\mathbf{\Omega}_2$.

$$\alpha h(\mathbf{x}) + \beta\frac{\partial h(\mathbf{x})}{\partial \mathbf{n}} = C_0, \quad \mathbf{x} \in \partial\mathbf{\Omega}_3 \qquad \textit{Robbins} \tag{1.11}$$

where C_0 is a specified function value along the boundary segment $\partial\mathbf{\Omega}$, and α and β are specified functions. Since the entire boundary must be defined by a boundary condition, the following relationship holds:

$$\partial\mathbf{\Omega} = \partial\mathbf{\Omega}_1 + \partial\mathbf{\Omega}_2 + \partial\mathbf{\Omega}_3 \tag{1.12}$$

Equation 1.9 is used to specify the head along the boundary. The specified relationship can be a function of time if the time-dependent behavior is known. Tidal behavior of the water surface of an estuary could, for example, be described by this kind of boundary specification.

When combined with the hydraulic conductivity or, in the case of a two-dimensional areal model, transmissivity, equation 1.10 provides a statement of the flux. In other words,

$$\mathbf{q_n}(\mathbf{x}) = -\mathbf{K} \cdot \frac{\partial h(\mathbf{x})}{\partial \mathbf{n}} \tag{1.13}$$

where $\mathbf{q_n}$ is the flux specified across the boundary $\partial\mathbf{\Omega}_2$.

The Robbins, or third-type boundary condition specified in equation 1.11 is used to describe what is called a **leakage condition** when used in a three-dimensional model. The Robbins boundary condition can also be used to represent a physical boundary at a long distance from a model boundary when this condition is used in a two- or three-dimensional model setting. The form used for both of these conditions is

$$\mathbf{K} \cdot \frac{\partial h(\mathbf{x})}{\partial \mathbf{n}} = \kappa(h_0(\mathbf{x}) - h(\mathbf{x})) \tag{1.14}$$

where h_0 **is the head external to the model and the head difference is taken in the direction n**. For example, in the case of a leakage situation, h_0 might be the **elevation in a lake** located above the top of the model. In this case the coefficient κ has the meaning of the resistivity to flow across the boundary that separates the water body from the aquifer. Alternatively, h_0 could be considered as the **head in the aquifer at some distance from the boundary**. In that case, the coefficient κ would be used as a surrogate for the distance to the location of h_0 relative to $\partial\mathbf{\Omega}_3$, that is, $\kappa \equiv K/L$, where L is the distance from the model boundary to the real boundary where h_0 is observed.

1.5.3 Stresses

As noted earlier, the stresses in a groundwater flow model are generally those associated with either net infiltration from rainfall or with pumping. The flow equation can be modified to account expressly for these two quite different stress conditions by rewriting the term Q in equation 1.7 as

$$Q = I + \delta(\mathbf{x} - \mathbf{x}_i)W \tag{1.15}$$

where I is **net infiltration** from precipitation or other sources, such as leaking sewers, and W_i is **discharge or recharge** at the well located at $\mathbf{x}_i$. The symbol $\delta(\mathbf{x} - \mathbf{x}_i)$ is the **Dirac delta function** and is used in finite-element analyses to locate the well pumpage W_i at point $\mathbf{x}_i$.

Net infiltration is generally determined using one of a series of formulae developed in soil physics and agronomy. Application of these formulae normally requires a knowledge of various factors, such as wind speed, plant growth, air temperature, soil characteristics, and precipitation. Lacking this information, it is often possible to get typical infiltration values for an area through local agricultural organizations.

Municipal well discharge and recharge may be available through records kept by local water supply agencies. Records on industrial use may also be available. However, for domestic wells, such information is not likely to be available and an estimate based on the demands of a typical family may have to suffice. Particularly in the case of municipal wells, the **time history of pumping** is generally important. In many areas of the country, **seasonal demands** for water can result in significantly different pumping patterns at different times of the year. These pumping patterns can have an important effect on groundwater flow directions, and especially on the movement of contaminants. As is the case for the parameters and boundary conditions described above, in actual practice **information on stresses will be required only at the nodes of the numerical mesh**.

1.6 WATER-TABLE CONDITION

1.6.1 Near-Surface Aquifer Zone

The **water table** can be simulated either as a sharp interface or as a variably saturated zone in which air and water coexist. While the variably saturated formulation is the more accurate representation of the water table, the sharp interface approximation is often used for historical reasons and because it is less complex. In the following we discuss, compare, and contrast the two approaches. We revisit these concepts in Section 1.7.2 when we talk about the dimensionality of models.

The water table is a somewhat ambiguous concept. It often means different things to different people. To understand how we will be using the concept in this book, consider the information presented in Figure 1.30. The rectangle on the left is a segment of soil observed in cross section. The section is divided into two parts by the water table. Below the water table exists the saturated zone, in which all pores are saturated with water. Above the water table is the **vadose zone**, in which water and air coexist. Between the water table and the **phreatic surface** (the zero-pressure surface) is the **capillary fringe**. In the capillary fringe the pores are filled with water, but the water is held in place by **capillary forces**.

To enhance our understanding of the physical processes active in the neighborhood of the water table, we now consider the diagram in the right-hand-side panel of Figure 1.30. On the horizontal axis is plotted the water saturation. A saturation of 1 means that all pores are occupied by water; the only air present is in solution. On the vertical axis is plotted the elevation relative to a datum identified as zero at the phreatic surface. This axis is also coincident with those that describe the water pressure and the hydraulic head (these are not shown). In a static system, the elevation, head, and water pressure are interrelated.

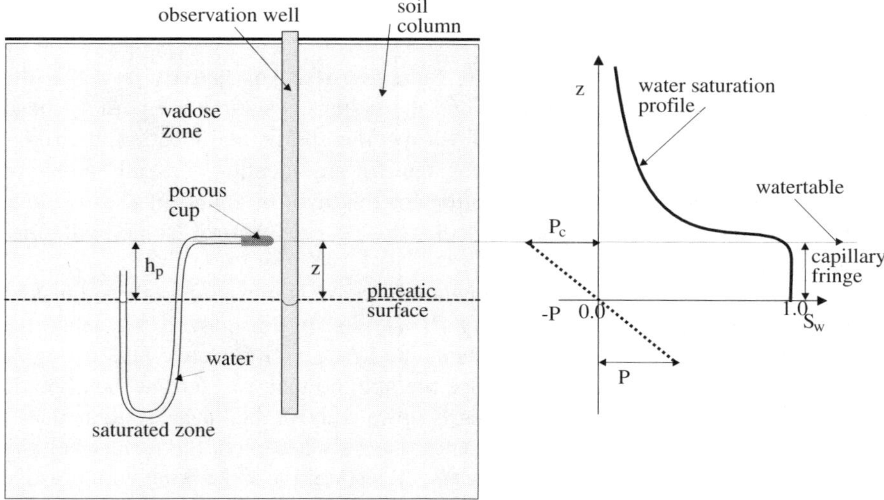

FIGURE 1.30. Definition sketch of components of the shallow-aquifer environment. The pressure head $h_p = P/\rho g$ and the balancing elevation head $h_e = z$. In a static system, changes in these two quantities at a specific horizontal location must balance.

The water-saturation profile is a description of the change in **water saturation** S_w with elevation z above the phreatic surface. In the context of our discussion, saturation is defined as the ratio of the volume of water to the volume of pore space in a reference volume of soil; it is often represented as a percentage. Note that at the phreatic surface, the saturation is 1; that is, all pore spaces are filled with water. For a distance (identified as the capillary fringe) above the phreatic surface, the saturation remains unity. At higher elevations the saturation decreases rapidly and approaches an asymptotic value. In other words, the water saturation approaches a lower limit with an increase in elevation. This lower limit is called the **irreducible saturation**. Saturation values below the irreducible saturation are achievable only through evaporation of water.

Because water above the phreatic surface is held in place via capillary forces, it is in tension. The water pressure in the capillary zone is below atmospheric. Since atmospheric pressure is normally denoted as zero, water in the capillary zone is under negative pressure. The lower the saturation, the more negative the pressure. In fact, in a static system, the pressure change is proportional to the elevation above the phreatic surface. Similarly, water below the phreatic surface is above atmospheric pressure and is therefore considered as positive. Under static conditions the pressure below the phreatic surface increases linearly with depth.

The pressure behavior observed can be understood more clearly if one considers the concept of hydraulic head as introduced in equations 1.5 to 1.7. The hydraulic head can be expressed, under certain simplifying assumptions, as

$$h(\mathbf{x}, t) = \frac{P(\mathbf{x}, t)}{\rho g} + z \tag{1.16}$$

where P is the **fluid pressure**, ρ is the **fluid density**, g is **gravity**, and z is the **elevation**. If the fluid is homogeneous and the system is static, the hydraulic head h must be constant everywhere. Thus, for a specified horizontal location, the right-hand side of equation 1.16 must be a constant for any location vertically above or below the specified point. The justification for this may be found in *Darcy's law*, which states that in saturated porous media, flow is proportional to the hydraulic-head gradient.

If h is constant, then for each incremental increase in elevation z, there must be a corresponding decrease in the quantity $P(\mathbf{x}, t)/\rho g$. In this context z is called the **elevation head**, which we will denote as h_e, and $P(\mathbf{x}, t)/\rho g$ is called the **pressure head**, which we will denote as h_p. Since we assume that z is zero at the phreatic surface, the pressure head decreases linearly upward above this surface and increases linearly downward below it. While the latter is easily accepted, the former is more abstract. Working with negative pressures is not as intuitive as working with positive pressures. The diagonal line passing through the origin of the graph in the right-hand-side panel of Figure 1.30 describes the behavior of fluid pressure below, at, and above the phreatic surface and illustrates the relationship between elevation and head.

Measuring negative water pressure is more challenging than measuring positive water pressure. One commonly used approach is shown in Figure 1.30. The porous cup is designed especially to measure negative water pressures. The pore size of this cup and the materials from which it is made are such that water is transmitted through it preferentially to air. If the tube attached to the cup is filled with water initially, the water level will drop on the open side of the tube until the difference in elevation between this water level and that of the porous cup is equivalent to a negative pressure head, indicative of that occurring at the location of the cup. The negative pressure head should equal the positive elevation head at that point since the reference is zero elevation at zero pressure (the phreatic surface). Observation of Figure 1.30 reveals that the difference in elevation between the water level in the pressure-measuring device (**tensiometer**) and the height of the porous cup above the phreatic-surface reference datum are exactly the same. Thus the net head at the porous cup in this case is zero, as it is at the phreatic surface, and therefore there is no flow.

In summary, we observe that the zone above the phreatic surface is characterized by negative water pressure. In the capillary zone this negative pressure is identified with saturated conditions and, in general, no free-phase air is present. The boundary between the capillary zone and the variably saturated zone above is herein called the **water table**. Below the phreatic surface the pressure increases with depth and the porous medium remains saturated.

It is very important to note that the concept of hydraulic head is consistent, irrespective of whether the measuring point is in the saturated, capillary, or partially saturated zones. In developing the governing equations, this observation is important and helpful.

1.6.2 Sharp-Interface Approximation of the Water Table

The water table, as described above, corresponds to the surface that separates the saturated zone from the variably saturated zone. However, the water table is also defined in some quarters as the surface of zero pressure. Since, in either case, the water table is a reasonably well-defined surface, it is often approximated as such. The resulting sharp-interface approximation is the foundation upon which the sharp-interface analysis formulation of Section 1.7.2 is based. However, before considering such a simplifying assumption, we consider the more complete variably saturated formulation.

1.6.3 Variably Saturated Water-Table Formulation

The equations that describe the flow of water in the variably saturated (also sometimes called *unsaturated*) zone are similar in form to those that govern saturated flow. The appropriate volume-conservation equation that is analogous to that presented as equation 1.5 is

$$\nabla \cdot \mathbf{q} = -S_s \frac{\partial h}{\partial t} - \varepsilon(1 - S_r)\frac{dS_w}{dh}\frac{\partial h}{\partial t} - Q \tag{1.17}$$

where the coefficient dS_w/dh is the slope of the saturation pressure (in this case pressure head) curve found in Figure 1.30 and reproduced as Figure 1.31. The term $(1 - S_r)$ accounts for the fact that only soil with a saturation greater than the residual (S_r) is mobile and can be drained.

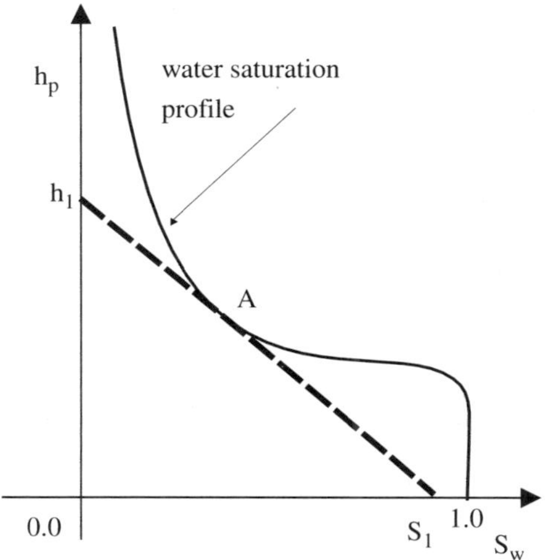

FIGURE 1.31. Pressure-saturation curve for variably saturated porous media.

Keep in mind that the pressure along the ordinate in this graph is increasing in a negative sense. The slope of the curve at point A is $(h_1 - 0)/(S_1 - 0)$ or $\Delta h_p/\Delta S_w$. The inverse of this ratio yields the desired coefficient, since the total head will change as a function of the pressure head given that the elevation z is not a function of time.

A physical interpretation of the coefficient dS_w/dh is best viewed in the context of the chain rule, that is, $\partial S_w/\partial t = (dS_w/dh)(\partial h/\partial t)$. From this expression it is evident that the additional term on the right-hand side of equation 1.17 is the rate of change of water volume in the control volume of soil due to **drainage** or **imbibition**. As is evident from Figure 1.31, the slope of the line used to represent dS_w/dh will change as one changes the value of the pressure head. Thus this coefficient (dS_w/dh) is a function of the hydraulic head. The product $(dS_w/dh)(\partial h/\partial t)$ is therefore nonlinear. Note that as one approaches total saturation, the slope of the pressure–saturation curve approaches infinity (dS_w/dh approaches zero). At this point the specific storage S_s is the dominant physical parameter providing water from storage. This is appropriate, since the porous medium is now saturated and the source of water is the elasticity of water or that of the soil matrix.

The Darcy flux $\mathbf{q}$ in equation 1.17 is given by a modified form of equation 1.6, that is,

$$\mathbf{q} = -\mathbf{K}_r(h) \cdot \nabla h \tag{1.18}$$

where the variably saturated hydraulic conductivity $\mathbf{K}_r$ is now a function of the hydraulic head h. The reason for this functional dependence of hydraulic conductivity on hydraulic head is found in the experimental relationship shown in Figure 1.32. The ordinate is relative permeability k_r, which ranges from zero to 1. The abscissa is again the degree of saturation. The relative permeability is related to the hydraulic conductivity via the relationship

$$\mathbf{K}_r(h) = \mathbf{K}k_r(h) \tag{1.19}$$

where $\mathbf{K}$ is the saturated hydraulic conductivity. The role of the relative permeability is therefore to scale the hydraulic conductivity to account for those pore spaces occupied by air. Air tends to block the flow of water. Note that the relative permeability curve vanishes at a saturation indicated by S_r. The parameter S_r is the **residual saturation** value, which is the saturation at which the water in different pores becomes disconnected. Below this saturation the water phase becomes immobile.

By combining Figures 1.31 and 1.32, we obtain Figure 1.33. The lower panel relates saturation to pressure head and the upper panel describes relative permeability in terms of saturation. Taking these two curves together, one obtains a relationship between relative permeability and pressure head as needed in equation 1.19.

Combination of equations 1.17 and 1.18 yields the equation describing the flow of water in both the saturated and partially saturated zones of an aquifer:

$$\nabla \cdot \mathbf{K}_r(h) \cdot \nabla h = \left[S_s + \varepsilon(1 - S_r)\frac{dS}{dh} \right] \frac{\partial h}{\partial t} + Q \tag{1.20}$$

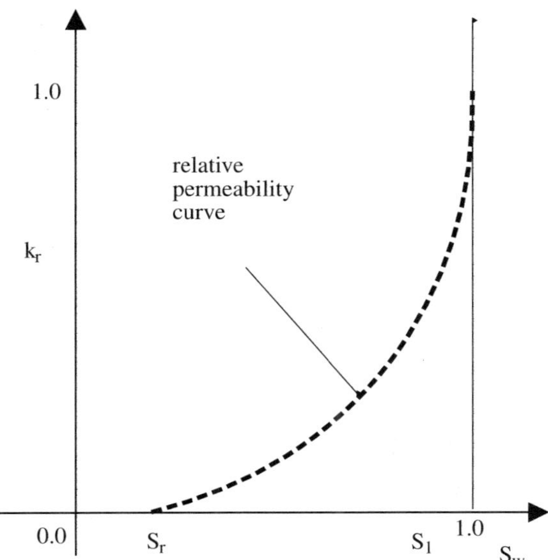

FIGURE 1.32. Relative-permeability curve. The parameter k_r is the relative permeability and S_r is the residual saturation.

1.6.4 Comparison of the Sharp-Interface and Variably Saturated Formulations

The question naturally arises as to when a sharp-interface formulation can be used and when a variably saturated simulation is required. An investigation of this question was made by Zhang [22]. The physical problem addressed was originally presented in Simunek and van Genucthen [23] and is illustrated in Figure 1.34. An aquifer composed of fine- and coarse-sand layers is being recharged by two constant-head reservoirs located at the top left and top right-hand corners of Figure 1.34. At the base of the aquifer there is a constant-head drain. The system is initially saturated and then allowed to drain to a steady-state condition. The governing equation is given by equation 1.20. The relationship between the saturation and the head used in this example is

$$S_w = \frac{1}{1 + (A|h_p|)^B}$$

and

$$B = \frac{1}{1 - C}$$

where h_p is the pressure head and A, B, and C are soil parameters presented in Table 1.6. The other required relationship is

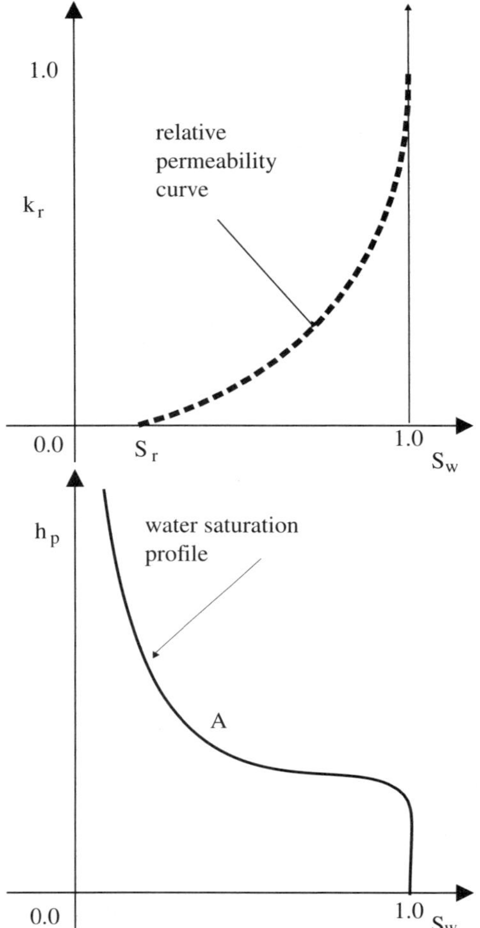

FIGURE 1.33. Combination of the saturation-pressure curve and relative-permeability curve.

$$\mathbf{K}_r = \mathbf{K} S_w^{1/2} \left[1 - \left(1 - S_w^{1/m_2} \right)^{m_2} \right]^2$$

where $m_2 = C$. The solution obtained using the unsaturated formulation presented above is given in Figure 1.36. In Figure 1.35 is presented the water-table evolution as calculated using the sharp-interface assumption.

A key to understanding the behavior observed is to realize that due to symmetry, only the left-hand side of the modeled region is being presented. It is also important to note that for clarity in presentation, the results plotted represent different times in Figure 1.35 than in Figure 1.36. For example, the first reporting period for the unsaturated flow results is after 0.001 day, whereas the first results presented for the sharp-

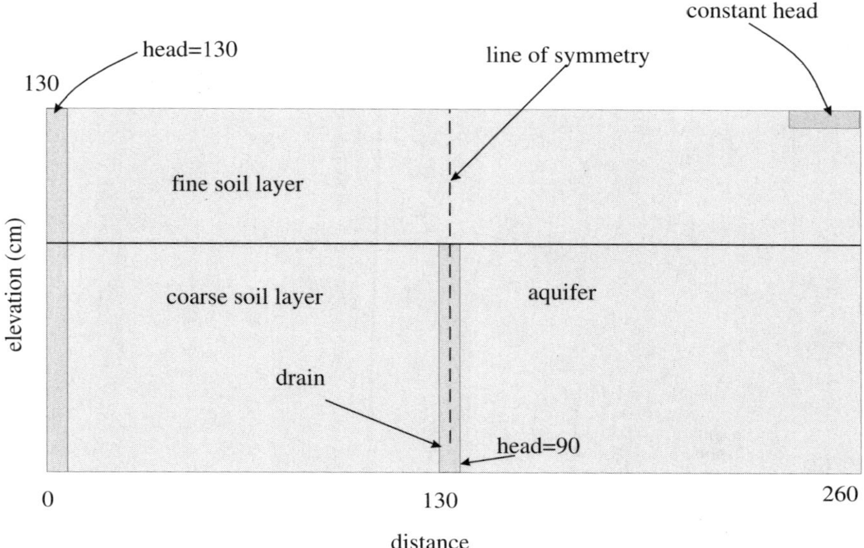

FIGURE 1.34. *Definition sketch for water-table problem. The constant head at the top of the aquifer is specified as 130 cm and at the drain the head is held at 90 cm.*

interface example are at an elapsed time of 0.0005 day. In comparing Figures 1.36 and 1.35 one observes that the steady-state solutions are similar. Thus for systems in which the steady-state solution is the primary simulation objective, the sharp-interface assumption is a reasonable one. On the other hand, there are significant differences between the transient sharp-interface and unsaturated-flow solutions. In this example, the sharp-interface representation of the water table moves at approximately twice the speed of that represented by solving the unsaturated-flow equations. The sharp interface formulation replaces the capillary zone by a sharp interface separating saturated and unsaturated zones. The saturation dependent parameters found in the unsaturated-flow solutions are essentially ignored and replaced by very simple representations. Given the difference in the mathematical–physical description of the system using the two approaches, the two solutions are remarkably similar for this problem.

TABLE 1.6. Parameters Used in the Unsaturated Representation of the Water-Table Example

Layer	Thickness (cm)	εS_r	εS_s	A	B	K (cm/day)
Fine soil	40	0.0001	0.399	0.0174	1.3757	29.8
Coarse soil	90	0.001	0.339	0.0139	1.6024	45.4

Source: Data from Zhang [22].

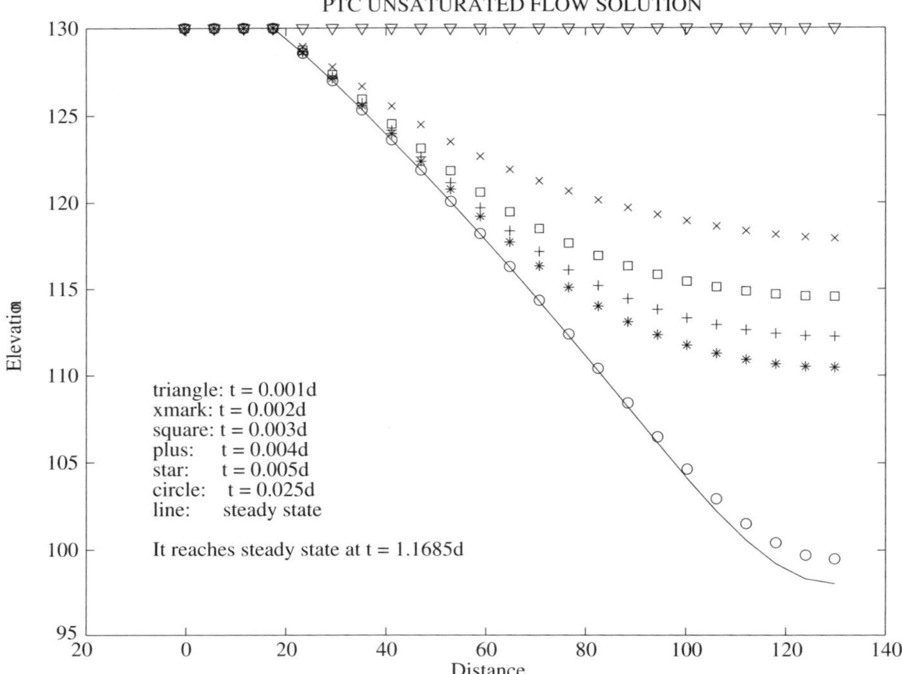

FIGURE 1.35. *Water-table elevation as computed using the sharp interface assumption in PTC.*

Although the water-table behavior is similar, the information content of the two solutions is quite different. The unsaturated-flow solution includes explicitly the flow of water in the unsaturated zone (and solutes if the transport equation is also solved). The sharp-interface solution considers only flow in the saturated zone, and no information is provided on movement in the unsaturated zone. On the other hand, the effort required to solve the unsaturated-flow equations is generally greater than in the case of the sharp-interface simulation.

1.7 PHYSICAL DIMENSIONS OF THE MODEL

In general, groundwater flow and transport simulations require a three-dimensional representation. In other words, groundwater flow is a three-dimensional process. In special circumstances one can simplify the simulation to require only a two-dimensional simulation. For example, when studying saltwater intrusion in a coastal aquifer, a **cross-sectional model** may be adequate. Similarly, when a vertically homogeneous aquifer is to be considered and any wells involved are nearly fully penetrating, a two-dimensional **horizontal (areal) model** may be adequate. However, in

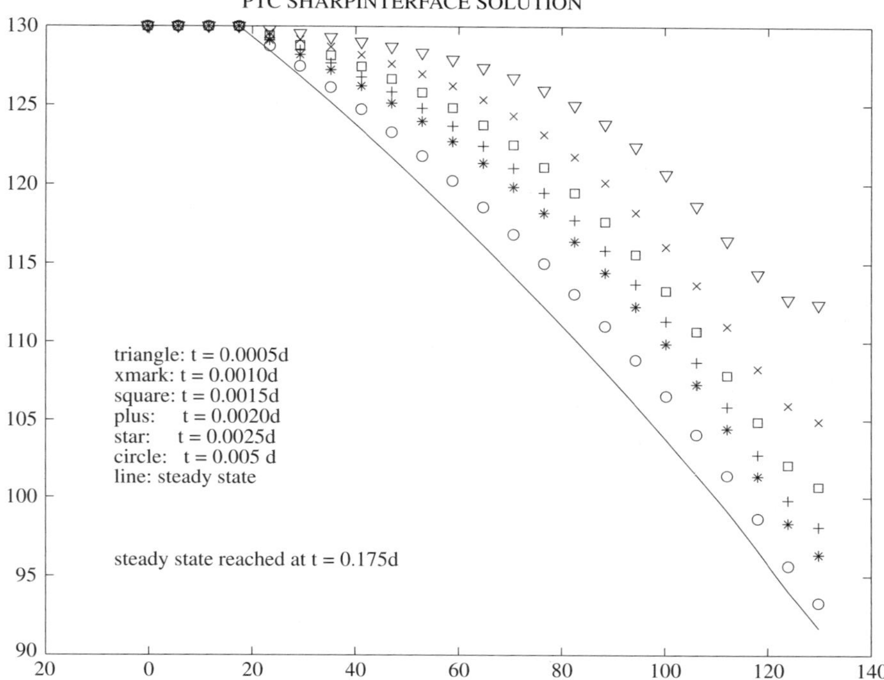

FIGURE 1.36. Water-table response obtained using the saturated-unsaturated flow model (from Zhang [22]).

general, it is necessary to justify that the dimension being neglected in simplifying a three-dimensional world to two dimensions can, in some sense, be disregarded.[21]

Even in this situation, it is necessary to be sure that the fundamental physics of the system are being preserved. For example, in a cross-sectional model, one is assuming that the behavior of **the groundwater system within any cross section along a line perpendicular to the section being considered (i.e., perpendicular to the paper) is the same as that for the cross section selected**. A very common error in this regard is to assume that groundwater flow in response to multiple wells can be represented in a Cartesian (x, z) cross section. This is not possible because wells generate radial flow patterns that cannot be represented, in general, in a Cartesian cross section. One can represent **flow to a single well in two dimensions**, but this **requires the use of a cylindrical (r, z) coordinate system**, not a Cartesian (x, z) coordinate system.[22]

[21] In actual fact, the dimension that is eliminated is not totally disregarded. Formally one is integrating over the neglected dimension and the missing dimension is being accommodated in this approximate sense.

[22] An exception to this statement is when a series of wells are located along a straight line such that an approximate line sink is created. This, of course, is not a likely scenario.

The most common simplification of the three-dimensional world of groundwater flow is to average over the vertical dimension to generate a two-dimensional areal model. Although this can be justified when flow is truly horizontal, one often hears the argument made that a two-dimensional model is desirable because too little is known about the geohydrological properties in the vertical dimension to justify modeling it. This is an incorrect concept. **Even when the aquifer is homogeneous vertically, flow in the third dimension may still be very important**. The correct question to ask is: Can the flow behavior in the vertical dimension be neglected without compromising the effectiveness of the model?

1.7.1 Vertical Integration of the Flow Equation

To understand what is involved in disregarding the vertical dimension, one must realize what is happening from the mathematical–physical point of view. This is best achieved by formally developing the areal two-dimensional model from the more general three-dimensional model. Let us begin with the flux form of the groundwater-flow equation,

$$\nabla \cdot \mathbf{q} = -S_s \frac{\partial h}{\partial t} - Q$$

and Darcy's law,

$$\mathbf{q} = -\mathbf{K} \cdot \nabla h \tag{1.21}$$

Consider the diagrammatic representation of the three-dimensional aquifer illustrated in Figure 1.37. The aquifer to be vertically integrated is illustrated in the figure. Performing the integration over the aquifer thickness, one obtains

$$\int_a^b \left(\nabla \cdot \mathbf{q} + S_s \frac{\partial h}{\partial t} + Q \right) dz = 0 \tag{1.22}$$

where a is the **top of the aquifer** and b is the **lower surface of the aquifer**. Application of **Leibnitz's rule** for differentiation of an integral,

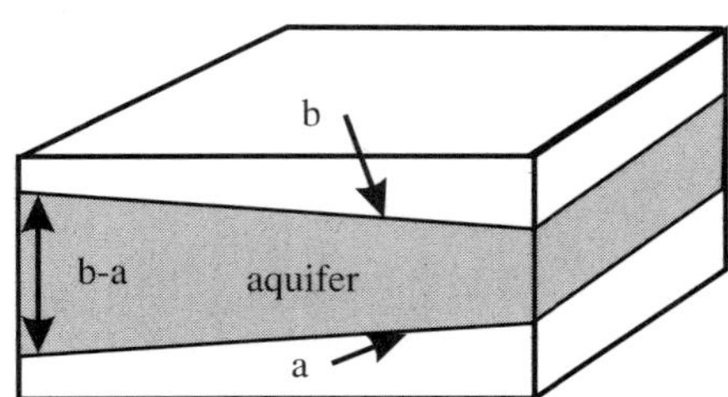

FIGURE 1.37. Diagrammatic representation for vertical integration.

$$\nabla \cdot \int_a^b \mathbf{q}(\mathbf{x})\,dz = \int_a^b \nabla \cdot \mathbf{q}(\mathbf{x})\,dz + \mathbf{q}(b) \cdot \nabla b - \mathbf{q}(a) \cdot \nabla a \tag{1.23}$$

to equation 1.22 yields

$$\int_a^b \left(\frac{\partial q_x}{\partial x} + \frac{\partial q_y}{\partial y} + \frac{\partial q_z}{\partial z} \right) dz = \frac{\partial}{\partial x} \int_a^b q_x\,dz - q_x|_b \frac{\partial b}{\partial x} + q_x|_a \frac{\partial a}{\partial x}$$
$$+ \frac{\partial}{\partial y} \int_a^b q_y\,dz - q_y|_b \frac{\partial b}{\partial y} + q_y|_a \frac{\partial a}{\partial y} + q_z|_a^b \tag{1.24}$$

or

$$\int_a^b \nabla \cdot \mathbf{q}\,dz = \nabla_{xy} \cdot \int_a^b \mathbf{q}_{xy}\,dz - \mathbf{q}_{xy}|_b \cdot \nabla_{xy} b + \mathbf{q}_{xy}|_a \cdot \nabla_{xy} a + q_z|_b - q_z|_a \tag{1.25}$$

where

$$\nabla_{xy}(\cdot) \equiv \frac{\partial(\cdot)}{\partial x}\mathbf{i} + \frac{\partial(\cdot)}{\partial y}\mathbf{j}$$

The **time derivative term** is treated in a similar way, that is,

$$S_s \int_a^b \frac{\partial h}{\partial t}\,dz = S_s \frac{\partial}{\partial t} \int_a^b h\,dz - S_s h\,|_b \frac{\partial b}{\partial t} + S_s h|_a \frac{\partial a}{\partial t} \tag{1.26}$$

From the general form of Darcy's law (equation 1.21), we obtain

$$\mathbf{q} = -\mathbf{K} \cdot \nabla h \tag{1.27}$$

Thus we obtain, using Leibnitz's rule,

$$\int_a^b \mathbf{q}\,dz = -\mathbf{K} \cdot \int_a^b \nabla h\,dz \tag{1.28}$$

$$= -\mathbf{K} \cdot \left(\nabla_{xy} \int_a^b h\,dz - h|_b \nabla_{xy} b + h|_a \nabla_{xy} a + h|_b \mathbf{k} - h|_a \mathbf{k} \right) \tag{1.29}$$

where $\mathbf{k}$ is the unit vector in the z-coordinate direction.

The combination of equations 1.25, 1.26, 1.27, and 1.28, and the assumption that $h|_b \simeq h|_a$, yields

$$- \nabla_{xy} \cdot \mathbf{K} \cdot \left(\nabla_{xy} \int_a^b h\,dz - h|_b \nabla_{xy} b + h|_a \nabla_{xy} a \right)$$
$$- \mathbf{q}_{xy}|_b \cdot \nabla_{xy} b + \mathbf{q}_{xy}|_a \cdot \nabla_{xy} a + q_z|_b - q_z|_a$$
$$+ S_s \frac{\partial}{\partial t} \int_a^b h\,dz - S_s h|_b \frac{\partial b}{\partial t} + S_s h|_a \frac{\partial a}{\partial t} + \int_a^b Q\,dz = 0 \tag{1.30}$$

Let us define the following averages:

$$\overline{h} = \frac{1}{l}\int_a^b h(\mathbf{x})\,dz$$

$$\overline{Q} = \frac{1}{l}\int_a^b Q(\mathbf{x})\,dz$$

where $l \equiv b - a$.[23] Assuming that $h|_a \simeq h|_b \simeq \overline{h}$, substitution of this definition into 1.30 yields

$$-\nabla_{xy}\cdot\mathbf{K}\cdot(\nabla_{xy}l\overline{h} - \overline{h}\nabla(b-a)) - \mathbf{q}_{xy}|_b\cdot\nabla_{xy}b + \mathbf{q}_{xy}|_a\cdot\nabla_{xy}a + q_z|_b - q_z|_a + S_s\frac{\partial}{\partial t}l\overline{h} - S_s\overline{h}\frac{\partial}{\partial t}(b-a) + \overline{Q}l = 0 \tag{1.31}$$

which, upon expansion of the derivatives, simplifies to

$$\nabla_{xy}\cdot\mathbf{T}_{xy}\cdot\nabla_{xy}\overline{h} = S\frac{\partial\overline{h}}{\partial t} - \mathbf{q}_{xy}|_b\cdot\nabla_{xy}b + \mathbf{q}_{xy}|_a\cdot\nabla_{xy}a + q_z|_b - q_z|_a + \overline{Q}l \tag{1.32}$$

where the **storage coefficient** $S \equiv S_s l$ and, as mentioned earlier, the **transmissivity** $\mathbf{T} \equiv l\mathbf{K}$.

Defining the flux through the top of the aquifer as q_T and that through the bottom as q_B, and defining the average flux being added to the aquifer as $q \equiv \overline{Q}l$, one obtains

$$\nabla_{xy}\cdot\mathbf{T}\cdot\nabla_{xy}\overline{h} = S\frac{\partial\overline{h}}{\partial t} + q_T + q_B + q \tag{1.33}$$

where $q_T = -\mathbf{q}_{xy}|_b\cdot\nabla_{xy}b + q_z|_b$ and $q_B = \mathbf{q}_{xy}|_a\cdot\nabla_{xy}a - q_z|_a$.

Note that the vertically averaged head appears in this equation and the head on the top and the bottom of the aquifer have been assumed equal to this average. Thus one can only justify using the areal two-dimensional form of the flow equation when the average head is a good representation of what is found in the aquifer and is approximately constant. In other words, **in the presence of significant vertical gradients, the areal two-dimensional form of the groundwater flow equation is not appropriate**.

1.7.2 Free-Surface Condition

The analysis above assumes a confined aquifer, that is an aquifer wherein there exist relatively impermeable geological units above and below the reservoir such that the reservoir remains totally saturated at all times. Let us now assume that the aquifer

[23] Note that $l = l(\mathbf{x})$ and the notation has been simplified for clarity in presentation.

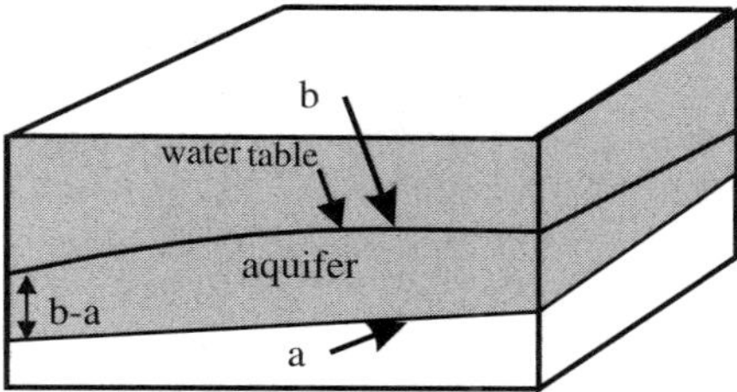

FIGURE 1.38. *Diagrammatic representation of the water table in an unconfined aquifer.*

is unconfined, that is, the reservoir contains the water table (Figure 1.38). Then the following analysis is relevant.

Define the **geometry of the free surface (water table) as** $F = F(\mathbf{x}, t)$. The **requirement that a particle on the free surface stay on the free surface is** $DF/Dt = 0$ where $D(\cdot)/Dt \equiv \partial(\cdot)/\partial t + \mathbf{v} \cdot \nabla(\cdot)$ is defined as the **substantial derivative**.

Consider the geometry of the upper surface as defined by $F = z - b(x, y, t) = 0$. Then we have from the condition for a free surface

$$\frac{DF}{Dt} = \frac{D}{Dt}(z - b) = \left(-\frac{\partial b}{\partial t} - \mathbf{v}_{xy} \cdot \nabla_{xy} b + v_z\right)\Big|_b = 0 \tag{1.34}$$

This statement can be interpreted to mean that a particle defining the free surface will remain on the free surface. It does not mean that a water molecule cannot pass through the locus of points defining the free surface. Indeed, in general, it can and will.

Multiplication of equation 1.34 by θ and subsequent subtraction of the result from equation 1.33 yields

$$\nabla_{xy} \cdot \mathbf{K}l(h) \cdot \nabla \overline{h} = S_s l(h) \frac{\partial \overline{h}}{\partial t} + \theta \frac{\partial b}{\partial t} + q_T + q_B + q \tag{1.35}$$

where $q_T = -\mathbf{q}_{xy}|_b \cdot \nabla_{xy} b + \theta \mathbf{v}_{xy} \cdot \nabla_{xy} b + (q_z - \theta v_z)|_b$ represents the net flow out of the aquifer across the water table. Note that the coefficients $\mathbf{K}l$ and $S_s l$ are now a function of the solution h, and therefore the **partial-differential equation is nonlinear**.

Tucson Example

The areal extent of the model has been considered in Section 1.2. The importance of considering discretization of the vertical dimension remains to be considered.

While the pumping pattern in the Tucson area is such that wells are often completed in the deep aquifer, much of the contamination resides in the shallow aquifer. In the absence of additional information, the pumping pattern would support the use of a three-dimensional model because of the induced vertical gradient. However, the existence of a **confining bed** that appears to have impeded movement of contaminants from the upper to the lower aquifer suggests that contaminant transport is largely two-dimensional in the areal plane and occurs primarily in the upper aquifer. But there is a complicating factor. The confining bed **pinches out** (vanishes) as one moves to the north–northeast in the study area. Thus the lower and upper units combine to form a single aquifer as one approaches the northernmost limits of the study area.

Given the complexity of this situation, one would be advised to consider a three-dimensional representation of this aquifer. This recommendation is predicated on the assumption of the availability of computer facilities that would permit prompt processing of three-dimensional model input.

1.8 MODEL SIZE

In the definition of the size of a model one must consider both the areal and vertical dimensions. As indicated in Section 1.2, the areal dimension is generally determined by four factors:

1. The **anticipated maximum extent of the response of stresses** to be imposed on the model is the first factor. In the case of a flow model, the important stresses are pumping stresses and the anticipated response is the expected resulting **cone of depression** (or mounding in the case of a recharge well).
2. The maximum areal extent of the model is defined in part by the availability of geohydrological boundary conditions. The various boundary conditions discussed above may define the areal extent of the model by virtue of their location. Sometimes, however, no suitable geohydrological boundary conditions are available. When this situation arises it is necessary to use other forms of boundary conditions, such as head values specified along a line determined from water levels in wells (Dirichlet conditions) or, in some instances, groundwater contour maps. Another alternative when faced with a lack of geohydrological boundary conditions to define a model is to use the third-type or Robbins boundary condition described above. As mentioned earlier, this can be used to locate the prototype boundary condition of the model sufficiently far from the model itself as to make the influence of the boundary condition relatively unimportant in the model domain.
3. The **top and bottom of a model** are normally defined by a relatively impermeable geological horizon. In the case of an aquifer located in an alluvial valley, for example, the relatively impermeable rock into which the valley is carved may act as the impermeable base for the aquifer (for that matter it might also form the impermeable sides of the model as well).
4. The **practical limitations imposed by the available computer facilities** may create constraints on the size of the model. As discussed in a later section, the amount of computational effort required to simulate a groundwater flow problem increases, in general, as the square of the number of nodes in the model. This number, in turn, is directly dependent on the level of discretization

used to model an area and the size of the area being modeled. Thus the size of the model is often limited by the size and speed of the computing platform available.

In summary, the physical attributes of the Tucson site suggest that a transient three-dimensional model would be appropriate for addressing most of the questions that are likely to be raised in the investigation and design process. However, if computer facilities are limited, an areal two-dimensional representation may be warranted.

1.9 MODEL DISCRETIZATION

1.9.1 Finite-Difference Approximations

In moving from a model based on partial-differential equations to one based on discrete equations, a numerical error is generated. In the case of a finite-difference model, the error is that identified with a Taylor's series approximation to a derivative. For example, in the case of $f = f(x)$,

$$f(x_0 + \Delta x) = f(x_0) + \left.\frac{df}{dx}\right|_{x_0} \cdot \Delta x + \frac{1}{2!}\left.\frac{d^2 f}{dx^2}\right|_{x_0} \cdot (\Delta x)^2 + \cdots \tag{1.36}$$

where Δx is an increment in the independent variable x (see, Figure 1.39, where a two-dimensional discretization is illustrated).

From equation 1.36 it is evident that the first derivative at x is given by

$$\left.\frac{df}{dx}\right|_{x_0} = \frac{f(x_0 + \Delta x) - f(x_0)}{\Delta x} - \frac{1}{2!}\left.\frac{d^2 f}{dx^2}\right|_{x_0} \cdot (\Delta x) - O(\Delta x)^2 \tag{1.37}$$

Thus the derivative df/dx evaluated at the location x can be **approximated** by $[f(x_0+\Delta x)-f(x_0)]/\Delta x$ with an error whose first term is $\frac{1}{2!}d^2 f/dx^2|_{x_0}\cdot(\Delta x)$. Con-

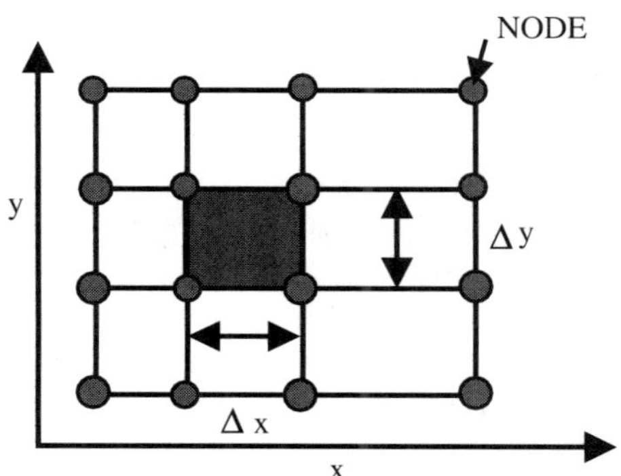

FIGURE 1.39. A two-dimensional discretization.

sequently, wherever the second derivative in f is large, or wherever Δx is large, so will be the error in the approximation of $df/dx|_{x_0}$, attributable to the first truncated term in the Taylor series approximation. As a result, if $d^2 f/dx^2|_{x_0}$ is large, it is necessary to have a very small value of Δx to assure that the error in the approximation of the derivative is small. A similar argument can be made for the approximation of the second derivative appearing in the flow equation. In this case, however, the error is proportional to $(\Delta x)^2$. It is clear at this point that the accuracy of the numerical approximation is dependent upon both the size of the discretization increment Δx and the behavior of the functions being approximated, in this case $f(x)$, and its derivatives.

In a practical sense, this observation means that in areas where there is a groundwater gradient that is changing rapidly, the spatial increments must be particularly small. Such situations occur around pumping or injection wells, in other areas of significant imposed stress, and in locations where the hydraulic conductivity changes abruptly.

1.9.2 Finite-Element Approximations

In the case of **finite-element approximations**, the formulation is formally very different, but the basic concept is the same. In finite elements, the unknown function, in the steady-state goundwater flow case $h(\mathbf{x})$, is approximated in one space dimension using the finite series

$$h(x) \simeq \hat{h}(x) = \sum_{i=1}^{I} h_i \phi(x)_i \tag{1.38}$$

where $h_i, i = 1, \ldots, I$ **are constants** (which turn out to be the values of the hydraulic head at the finite-element nodes). The **basis functions** $\phi(x)_i$ are normally chosen to be Lagrange polynomials of degree less than four. The lower the degree, the larger the truncation error of the approximation. **When linear basis functions are used (see equations 1.63 and 1.64 and Figure 1.43), the truncation error is generally of the same order as that encountered with standard finite-difference methods**. Thus, as in the case of finite-difference methods, smaller discretization is needed when the unknown function, in this case the hydraulic head, changes rapidly.

1.9.3 Two-Space Dimensional Approximations

As indicated above, whether finite-difference or finite-element methods are used on the same grid, theoretically they have approximately the same accuracy for the groundwater-flow equation. However, **finite-difference methods are normally used only on rectangular grids**, whereas **finite-element methods also can readily be formulated on triangles, or even on deformed rectangles with curved sides**. The latter elements are often called **isoparametric finite elements**.

The **importance of nonrectangular elements** lies in their ability to represent irregular boundaries, irregular areas of a different hydraulic conductivity, or denser

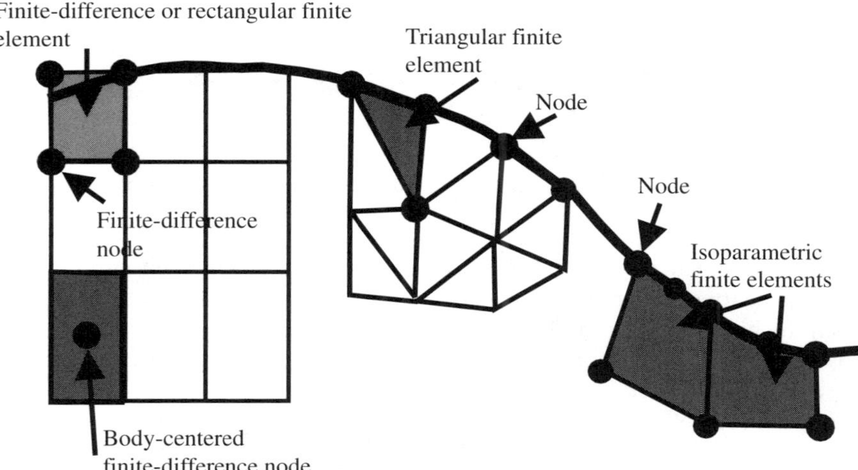

FIGURE 1.40. *Finite-element, finite-difference, and body-centered finite-difference (integrated finite difference) meshes.*

element meshes in the neighborhood of large stresses such as those found around wells. Figure 1.40 illustrates how finite-difference, finite-element, and isoparametric finite-element methods can be used to represent an irregular boundary. It is evident in this figure that the finite-element approach will better represent the curved boundary. You will note in Figure 1.40 that two kinds of finite-difference subspaces are shown. One has **nodes on the corners** of the element and the other has **nodes in the center (also described in the literature as integrated finite-difference or finite-volume elements)**. To understand the difference between these two kinds of finite-difference elements, remember that the node is the reference location at which the finite-difference approximation for the derivative is written. Consider the one-dimensional example of Figure 1.41.

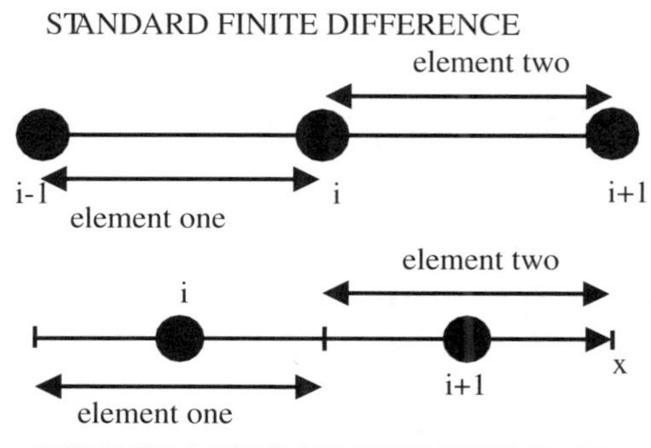

FIGURE 1.41. *One-dimensional finite-difference discretization.*

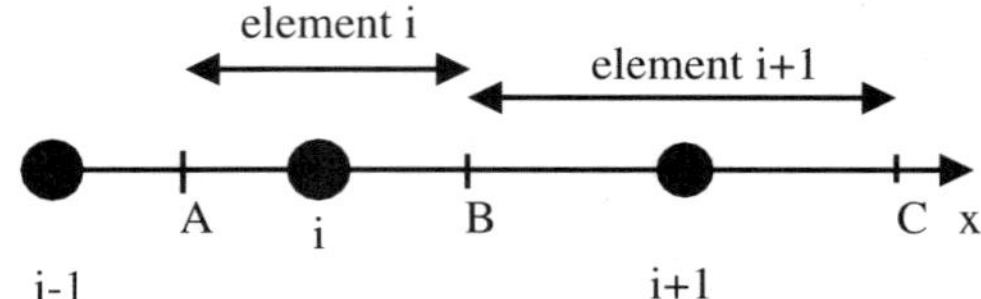

FIGURE 1.42. *One-dimensional finite-difference mesh with variable spacing.*

An approximation of second derivatives, such as found in the flow equations, will be the same irrespective of whether the mesh is standard or body centered. However, when the body-centered mesh is used with variable mesh spacing, that is, $\Delta x_i \neq \Delta x_{i+1}$, there is a conceptual advantage to the body-centered approach. Consider in this regard Figure 1.42.

It is conceptually attractive to think of the element for node i as having the node in the center. In this way, one can think of the two-dimensional version of the block-centered mesh as being made up of boxes with the nodes in the middle. This image allows one to accept quickly the idea of writing a **mass balance** at the edges of the boxes: in other words, to recognize that the mass flowing out of all sides must equal the amount of mass lost from storage plus that provided by sources. The fluxes at the sides of the boxes are described by Darcy's law. Thus by approximating the fluxes in Darcy's law, one is also creating the mass-balance equation for groundwater-flow.

In the case of finite-element methods, one has no choice regarding the location of the nodes vis-à-vis the element boundaries. As will be seen later, nodes are always located at the corners or on the edges of finite elements. However, a variant on the finite-element method, called the **finite-volume method (the same concept mentioned earlier within the context of body-centered finite-difference methods)**, allows one to create element configurations in finite elements similar to those found for body-centered finite differences (or finite volumes).

There is no free lunch. While the finite-element method and other nonrectangular mesh techniques provide **increased flexibility** in the use of elements, they also generally require **more computational effort** to solve the resulting set of equations given the same number of nodes. This is due to the fact that the matrices generated by the finite-element method are less regularly structured than those generated by the finite-difference method. As a consequence, many computationally efficient **algebraic equation solvers** that are applicable to finite-difference matrices are not effective on finite-element matrices. More information on this topic is given in Section 1.16.

1.10 FINITE-DIFFERENCE APPROXIMATION TO THE FLOW EQUATION

To this point, it has not been necessary for us to know in much detail about the numerical approximation of the equations that we are using. However, before we can enter the boundary conditions, parameters, and stresses for our model, it is necessary to decide upon the numerical procedure that is going to be used. The model that is used extensively in this book uses both finite-difference and finite-element concepts, so we next describe both briefly.

Consider the flow equation presented earlier as equation 1.7:

$$\nabla \cdot \mathbf{K} \cdot \nabla h = S_s \frac{\partial h}{\partial t} + Q$$

This equation is second order in space and first order in time. Thus it is necessary to approximate both first- and second-order derivatives. A complication is added by the possibility of a spatially variable hydraulic conductivity $\mathbf{K} = \mathbf{K}(\mathbf{x})$. Indeed, in the most general case, as noted earlier, $\mathbf{K}$ is a tensor, as indicated by its boldface type. In our work, and in virtually all practical applications, we will assume that only the diagonal elements of the hydraulic conductivity tensor are nonzero. This assumption conveniently eliminates the cross derivatives that would otherwise appear in the flow equation. However, there is no theoretical reason why the cross-derivatives cannot be accommodated.

Let us consider a finite-difference formulation built around the template in Figure 1.42. There are several ways to formulate this equation. We begin by writing the expression for the flux at the interelement boundaries A and B:

$$q_A = \left[-K_{xx} \frac{dh}{dx} \right]_A \simeq -K_{xx} \Big|_A \frac{h_i - h_{i-1}}{\Delta x_A} \tag{1.39}$$

$$q_B = \left[-K_{xx} \frac{dh}{dx} \right]_B \simeq -K_{xx} \Big|_B \frac{h_{i+1} - h_i}{\Delta x_B} \tag{1.40}$$

We now approximate the divergence of velocity as

$$\left. \frac{dq}{dx} \right|_i \simeq \frac{q_B - q_A}{\Delta x_i} \tag{1.41}$$

Substitution of equations 1.39 and 1.40 into 1.41 yields

$$\left. \frac{dq}{dx} \right|_i = \frac{-K_{xx}|_B[(h_{i+1} - h_i)/\Delta x_B] + K_{xx}|_A[(h_i - h_{i-1})/\Delta x_A]}{\Delta x_i} \tag{1.42}$$

which is the finite-difference approximation to the left-hand side of equation 1.5. The question that remains to be answered is the definition of the terms $K_{xx}|_A$, $K_{xx}|_B$, Δx_A, and Δx_B.

Consider Darcy's law written on either side of point B in Figure 1.42. For clarity we drop the subscripts on the hydraulic conductivity. We then obtain

$$q|_{B_-} = -\frac{K_i}{\Delta x_i/2}(h_B - h_i) \tag{1.43}$$

$$q|_{B_+} = -\frac{K_{i+1}}{\Delta x_{i+1}/2}(h_{i+1} - h_B) \tag{1.44}$$

Now multiply from the left both sides of equations 1.43 and 1.44 by the same coefficient, that is,

$$\frac{K_{i+1}}{\Delta x_{i+1}/2} \cdot q|_{B_-} = -\frac{K_{i+1}}{\Delta x_{i+1}/2} \cdot \frac{K_i}{\Delta x_i/2}(h_B - h_i) \tag{1.45}$$

$$\frac{K_i}{\Delta x_i/2} \cdot q|_{B_+} = -\frac{K_i}{\Delta x_i/2} \cdot \frac{K_{i+1}}{\Delta x_{i+1}/2}(h_{i+1} - h_B) \tag{1.46}$$

Sum equations 1.45 and 1.46 to yield

$$\frac{K_{i+1}}{\Delta x_{i+1}/2} \cdot q|_{B_-} + \frac{K_i}{\Delta x_i/2} \cdot q|_{B_+} = -\frac{K_{i+1}}{\Delta x_{i+1}/2} \cdot \frac{K_i}{\Delta x_i/2}(h_B - h_i) - \frac{K_i}{\Delta x_i/2} \cdot \frac{K_{i+1}}{\Delta x_{i+1}/2}(h_{i+1} - h_B)$$

Since the flux at the point B must be the same from either direction, $q|_{B_-}$ must equal $q|_{B_+}$. Thus, equating these two fluxes, we obtain

$$q_B = \frac{[K_{i+1}/(\Delta x_{i+1}/2)][K_i/(\Delta x_i/2)]}{[K_{i+1}/(\Delta x_{i+1}/2)] + [K_i/(\Delta x_i/2)]}(h_{i+1} - h_i) \tag{1.47}$$

which illustrates that **the correct evaluation of the coefficients at i and $i+1$ associated with the interval between these two nodes is the harmonic mean,**[24] defined by

$$\frac{K_B}{\Delta x|_B} = \frac{[K_{i+1}/(\Delta x_{i+1}/2)][K_i/(\Delta x_i/2)]}{[K_{i+1}/(\Delta x_{i+1})/2] + [K_i/(\Delta x_i)/2]} \tag{1.48}$$

To write the groundwater flow equation in one dimension (which is all we will need in the formulation that is used in the software for this book), we must first approximate the time derivative that appears in equation 1.5. This requires the use of double subscripts, as shown in Figure 1.39. Thus we have $h\,(x_i, t_n) \equiv h_{i,n}$. We can now write, from equation 1.37,

$$\left.\frac{\partial h}{\partial t}\right|_{x_i t_{n+1}} = \frac{h_{i,n+1} - h_{i,n}}{\Delta t} + O(\Delta t) \tag{1.49}$$

Combining equations 1.42, 1.48, and 1.49, we arrive at our one-dimensional finite-difference approximation to the flow equation

$$\frac{-K|_B[(h_{i+1,n+1} - h_{i,n+1})/\Delta x_B] + K|_A[(h_{i,n+1} - h_{i-1,n+1})/\Delta x_A]}{\Delta x_i} = S_s\frac{h_{i,n+1} - h_{i,n}}{\Delta t} \tag{1.50}$$

where the coefficients evaluated at A and B are defined in equation 1.48.

[24]The harmonic mean of a and b is $1/(1/a + 1/b)$ or $(a \cdot b)/(a + b)$.

1.10.1 Model Boundary Conditions

Second-Type Boundary Conditions If a flow boundary condition is located at position i, the two nodes at $i-1$ and $i+1$ are involved; visualize

$$\left.\frac{dh}{dx}\right|_i = \frac{h_{i+1} - h_{i-1}}{2\Delta x} + O(\Delta x)^2 \tag{1.51}$$

where Δx is, once again, the length of the finite-difference increment in the x direction. Now consider the body-centered finite-difference case. If the boundary is located at i, one approximation would read

$$\left.\frac{dh}{dx}\right|_i \simeq \frac{h_{i+1} - h_i}{\Delta x} + O(\Delta x) \tag{1.52}$$

which is a first-order approximation and therefore is less accurate than that found in equation 1.51. However, if we imagine writing the derivative approximation at the interface between the two elements in the body-centered net shown in Figure 1.41, we have

$$\left.\frac{dh}{dx}\right|_{i+1/2} \simeq \frac{h_{i+1} - h_i}{\Delta x} + O(\Delta x)^2 \tag{1.53}$$

which is **second-order accurate** because the derivative is being approximated at a location midway between the two nodes i and $i+1$. Thus one observes that the manner in which a flux boundary condition is interpreted in terms of its location vis-à-vis that of existing nodes can dictate the accuracy of the boundary approximation. A similar sort of development can be used for leakage (third-type or Robbins) boundary conditions.

First-Type Boundary Conditions On the other hand, **constant head (first-type or Dirichlet)** boundary conditions are normally accommodated simply by replacing the unknown head value in the algebraic equations with a known head value. When the nodes are located on the boundaries, this does not introduce any additional error into the approximation. If, however, one imagines the boundary to be at $i+\frac{1}{2}$, an order of Δx error is committed if either the nodal value at i or at $i+\frac{1}{2}$ is used. An order of approximation of $(\Delta x)^2$ can be obtained by taking the arithmetic average of the two values, h_i and h_{i+1}, and setting the result equal to the specified value at location $i+\frac{1}{2}$. The main difficulty with the latter scheme is that the number of unknowns (i.e., the number of unknown values of h) is not decreased.

1.10.2 Model Initial Conditions

In a time-evolution problem such as that associated with the response of a groundwater system to new stresses, it is necessary to provide the model with information regarding the state of the system at the time the simulation begins. The need for this information is evident in the form of the groundwater equation. The existence of a first derivative in time indicates at once that an **initial-head state** is needed to provide a unique model solution. Thus, along with boundary conditions, it is necessary to provide a head value indicative of the initial state of the system at each node in the interior of the model.

The **initial conditions** associated with the boundary points have already been accommodated by the boundary conditions. In other words, the boundary conditions for the beginning of the simulation must, theoretically, be consistent with the boundary conditions. However, as we will see in the next paragraph, this may not be a major concern.

Fortunately, the groundwater **flow system** responds rapidly to new boundary conditions or new stresses, normally in terms of hours or days rather than years. As a result, the initial conditions are not of great importance in most groundwater problems. The system adjusts rapidly and the initial conditions are soon of little importance. Exceptions to this general rule exist when the groundwater system contains stratigraphic layers of low hydraulic conductivity or when the system is very large; that is, it has areal dimensions of tens of miles.

An important exception to the assumption that initial conditions are generally of relatively little importance occurs in the case of **multiple pumping periods**. When the pumping rate changes in the model, the new pumping rates must be introduced as new point boundary conditions and the model rerun for the period of time for which the new conditions apply. In this instance, **the initial state of the system for the new pumping rates is the final state of the system for the previous pumping rates**. The final head values for the system that were obtained for the previous pumping strategy are now introduced as the initial conditions for the new pumping campaign.

1.11 FINITE-ELEMENT APPROXIMATION TO THE FLOW EQUATION

The finite-element approximation of the flow equation is most easily formulated using the Galerkin method of weighted residuals. The point of departure is the finite-series approximation for the unknown parameter $h(\mathbf{x}, t)$ presented earlier, that is,

$$h(\mathbf{x}, t) = \hat{h}(\mathbf{x}, t) = \sum_{i=1}^{I} h(t)_i \phi(\mathbf{x})_i \tag{1.54}$$

where the coefficient h_i is now recognized as a function of time. The reason for making this coefficient time dependent rather than simply to take the alternative route of making $\phi(\mathbf{x})_i$ time dependent will become apparent shortly. As mentioned earlier, the basis function $\phi(\mathbf{x})_i$ will normally be a linear, or at most quadratic, function.

The method of weighted residuals can now be formulated. The first step is to substitute the approximating function $\hat{h}(\mathbf{x}, t)$ into the flow equation, that is,

$$\nabla \cdot \mathbf{K} \cdot \nabla \hat{h} - S_s \frac{\partial \hat{h}}{\partial t} - Q = R(\mathbf{x}, t) \tag{1.55}$$

where the residual $R(\mathbf{x}, t)$ is, in general, nonzero because the approximating function $\hat{h}(\mathbf{x}, t)$ does not exactly satisfy the governing equation.

The next step is to weight this residual $R(\mathbf{x}, t)$ by the basis function $\phi(\mathbf{x})_i$ and to set the integral of this product over the model domain to zero, that is,

$$\int_{\Omega} R(\mathbf{x}, t)\phi(\mathbf{x})_i \, d\Omega = 0, \qquad i = 1, \ldots, I \tag{1.56}$$

Notice that this does not guarantee that the residual is zero at any point in the domain but rather, that its weighted integral over the domain is zero. Substituting for $R(\mathbf{x}, t)$, one obtains

$$\int_{\Omega} \left(\nabla \cdot \mathbf{K} \cdot \nabla \hat{h} - S_s \frac{\partial \hat{h}}{\partial t} - Q \right) \cdot \phi(\mathbf{x})_i \, d\Omega = 0, \qquad i = 1, \ldots, I \tag{1.57}$$

The next step in the development is to apply Green's theorem (integration by parts in multiple dimensions) to equation 1.57. This yields

$$\int_{\Omega} \left[(-\mathbf{K} \cdot \nabla \hat{h}) \cdot \nabla \phi(\mathbf{x})_i - \left(S_s \frac{\partial \hat{h}}{\partial t} + Q \right) \phi(\mathbf{x})_i \right] d\Omega$$
$$+ \int_{\partial \Omega} \mathbf{K} \cdot \frac{\partial h}{\partial \mathbf{n}} \phi(\mathbf{x})_i \, dl = 0, \qquad i = 1, \ldots, I \tag{1.58}$$

Note that in equation 1.58 a new term, the last term on the left-hand side of this equation, has been introduced via Green's theorem: namely, the surface integral (or line integral for a two-dimensional problem). This term has the form of a flux term. Indeed, when this term is needed, it will be known (or can be evaluated) via a type 2 (Neumann or flux) boundary condition. To understand further how this is done, let us proceed a little further in the finite-element development by substituting $\hat{h}(\mathbf{x}, t) = \sum_{j=1}^{J} h(t)_j \phi(\mathbf{x})_j$ in equation 1.58. We obtain

$$\int_{\Omega} \left[\left\{ -\mathbf{K} \cdot \nabla \sum_{j=1}^{J} h(t)_j \phi(\mathbf{x})_j \right] \cdot \nabla \phi(\mathbf{x})_i \right.$$
$$\left. - \left[S_s \frac{\partial \sum_{j=1}^{J} h(t)_j}{\partial t} \phi(\mathbf{x})_j + Q \right] \phi(\mathbf{x})_i \right\} d\Omega$$
$$+ \int_{\partial \Omega} \mathbf{K} \cdot \frac{\partial h}{\partial \mathbf{n}} \phi(\mathbf{x})_i \, dl = 0, \qquad i = 1, \ldots, I \tag{1.59}$$

At this point one can see that equation 1.59 represents I equations. We will now utilize a finite-difference representation for the time derivative $\partial h(\mathbf{x}, t)/\partial t$. Thus we obtain

$$\left. \frac{\partial h(t_{n+1})}{\partial t} \right|_{x_i} = \left. \frac{h_{n+1} - h_n}{\Delta t} \right|_{x_i} + O(\Delta t) \tag{1.60}$$

which, when substituted into equation 1.59 yields

$$\int_{\Omega}\left[\left\{-\mathbf{K}\cdot\nabla\sum_{j=1}^{J}h_{n+1,j}\phi(\mathbf{x})_j\right]\cdot\nabla\phi(\mathbf{x})_i\right.$$

$$\left.-\left[S_s\sum_{j=1}^{J}\frac{h_{n+1,j}-h_{n,j}}{\Delta t}\phi(\mathbf{x})_j+Q\right]\phi(\mathbf{x})_i\right\}d\Omega$$

$$+\int_{\partial\Omega}\mathbf{K}\cdot\frac{\partial h}{\partial \mathbf{n}}\phi(\mathbf{x})_i\,dl=0,\qquad i=1,\ldots,I \tag{1.61}$$

One now has I algebraic equations in J unknowns. Thus, if $I = J$, as is the case in this formulation, it should be possible to solve equation 1.61 for the J unknown values of $h_{n+1,j}$ at each time step.

But what about the integrals in equation 1.61 you ask? Good question! The answer to this question constitutes the heart of the finite-element method.

Perhaps the easiest way to look at this integration process is to reduce the problem to one space dimension, much as we did for the finite-difference discussion presented above. Thus we have for equation 1.61

$$\int_{l}\left[\left(-K\frac{d}{dx}\sum_{j=1}^{J}h_{n+1,j}\phi(x)_j\right)\frac{d}{dx}\phi(x)_i\right.$$

$$\left.-\left(S_s\sum_{j=1}^{J}\frac{h_{n+1,j}-h_{n,j}}{\Delta t}\phi(x)_j+Q\right)\phi(x)_i\right]dx$$

$$+K\frac{\partial h}{\partial n}\phi(x)_i\Big|_{x_0}^{x_l}=0,\qquad i=1,\ldots,I \tag{1.62}$$

To proceed further, it is helpful to consider a specific form for the basis functions $\phi(x)_j$. Figure 1.43 shows **linear basis functions**, sometimes called **chapeau functions** because of their hatlike shape. The forms of these functions are

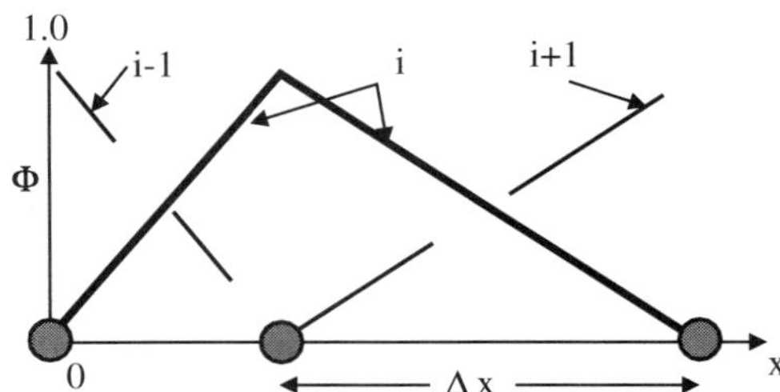

FIGURE 1.43. *Linear chapeau basis functions.*

$$\phi(x)_i = \frac{x - x_{i+1}}{x_i - x_{i+1}}, \qquad x_i \leq x \leq x_{i+1} \tag{1.63}$$

$$\phi(x)_i = \frac{x - x_{i-1}}{x_i - x_{i-1}}, \qquad x_{i-1} \leq x \leq x_i \tag{1.64}$$

With this structure in mind, it becomes evident that the integral in equation 1.62 is actually made up of E piecewise integrals. Thus we can rewrite equation 1.62 as

$$\sum_{e=1}^{E} \int_e \left[-K \frac{d}{dx} \sum_{j=1}^{J} h_{n+1,j} \phi(x)_j \right] \frac{d}{dx} \phi(x)_i$$
$$- \left[S_s \sum_{j=1}^{J} \frac{h_{n+1,j} - h_{n,j}}{\Delta t} \phi(x)_j + Q \right] \phi(x)_i dx$$
$$+K \frac{\partial h}{\partial n} \phi(x)_i |_{x_0}^{x_I} = 0, \qquad i = 1, \ldots, I \tag{1.65}$$

We now see that the typical integral is of the form

$$- K_e \cdot \int_e \left[\frac{d}{dx} \sum_{j=1}^{J} h_{n+1,j} \phi(x)_j \right] \frac{d}{dx} \phi(x)_i \, dx$$
$$- S_{s_e} \int_e \left[\sum_{j=1}^{J} \frac{h_{n+1,j} - h_{n,j}}{\Delta t} \phi(x)_j + Q \right] \phi(x)_i \, dx, \qquad i = 1, \ldots, I \tag{1.66}$$

where the parameters K and S_s are now assumed to be constant over each element e. Because the integrals to be evaluated involve integrands that contain the product of linear or constant functions, these integrations are easily done by computer. We visit this issue later.

1.11.1 Boundary Conditions

Let us now return to the question of boundary conditions.

Second-Type Boundary Conditions The last term in equation 1.65 contains embedded in it second-type (Neumann) boundary conditions. This fact is more evident when one realizes that the basis functions at the ends of the domain are either zero or 1. Thus this last term becomes

$$\mathbf{K} \cdot \frac{\partial h}{\partial \mathbf{n}} \quad \text{at} x = l \tag{1.67}$$

$$-\mathbf{K} \cdot \frac{\partial h}{\partial \mathbf{n}} \quad \text{at} x = 0 \tag{1.68}$$

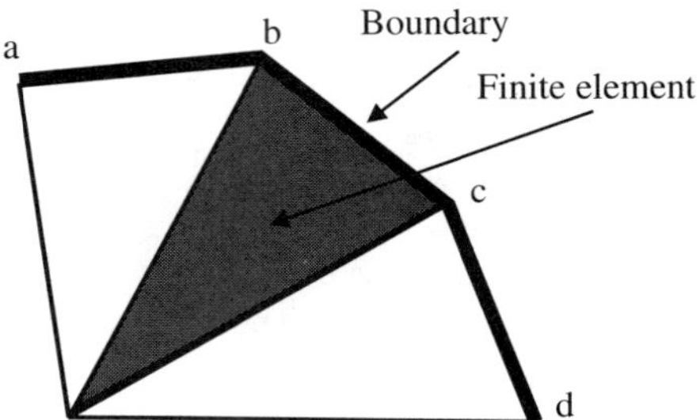

FIGURE 1.44. Definition of a domain boundary using triangular finite elements.

Equation 1.67 represents flow into the aquifer from location l. Thus simply by replacing this term with the appropriate value of the boundary condition incorporates this boundary condition as an integral part of the governing equation. This is considered to be a significant computational and conceptual benefit of the finite-element approach.

The extension of this concept to two space dimensions must now be addressed. Consider Figure 1.44. The line *abcd* in Figure 1.44 we will consider to be the boundary on a finite-element net, three triangles of which are indicated in this figure. The integral that constitutes the boundary condition in this case is given by

$$\int_{\partial\Omega} \mathbf{K} \cdot \frac{\partial h}{\partial \mathbf{n}} \phi(\mathbf{x})_i \, dl \tag{1.69}$$

which is found in equation 1.61. The appropriate integral in this specific case is given by

$$\int_{e=1}^{3} \mathbf{K} \cdot \frac{\partial h}{\partial \mathbf{n}} \phi(\mathbf{x})_i \, dl, \qquad i = a, b, c, d \tag{1.70}$$

where the functions $\phi(\mathbf{x})_i$ are given by equations 1.63 and 1.64 and illustrated in Figure 1.43. The result of this integration is the division of the product of the flux and the element length by 2, with the resulting value being assigned to each of the element nodes.

For example, if the flux along side a–b in Figure 1.44 was 5 ft^2/day per foot of boundary, the flux value allotted to node a due to the existence of this element would be $\frac{5}{2} \cdot (b - a)$, where $(b - a)$ is the line length. This number would replace the boundary term in equation 1.70 for $i = a$. For the case of $i = b$ there would be contributions from the two elements with common node b.

Third-Type Boundary Conditions In the case of a **third-type boundary condition**, one replaces the integral in equation 1.70 by an expression of the form

$\kappa(h - h_0)$, where h_0 is a specified function. In this case the unknown h appears in the boundary condition. This term must then be approximated using the finite-element basis functions, as shown in equation 1.54. The resulting unknown parameters in this expression must then be obtained along with the other unknown parameters $h_j(\mathbf{x}, t)$.

First-Type Boundary Conditions To accommodate a first-type (constant head) boundary condition, one should once again examine equation 1.54,

$$h(\mathbf{x}, t) = \hat{h}(\mathbf{x}, t) = \sum_{i=1}^{I} h(t)_i \phi(\mathbf{x})_i$$

Note that when a point $\mathbf{x}_i$ is the location of a node, the basis function $\phi(\mathbf{x})_i$ has a value of 1.0 and every other basis function has a value of zero. Thus at a node i, this equation becomes simply

$$h(\mathbf{x}_i, t) = \hat{h}(\mathbf{x}_i, t) = h(t)_i$$

Thus, by assigning to the parameter $h(t)_i$ a specific value, $h(t)_0$ say, the first type boundary condition is satisfied.

1.11.2 Initial Conditions

The specification of initial conditions is straightforward. The initial head value at each node in the system is given a value indicative of the initial state of the system. The protocol for entering this information into the model is analogous to that described in the Tucson example in Section 1.4.5. However, if observed values of hydraulic head are used as initial conditions, these values will almost certainly not represent an accurate solution to the groundwater flow equations. Thus, when the model calculations are initiated, the initial conditions will change in such a way as to satisfy the governing equations. The resulting behavior of the system may be quite unexpected as the system modeled attempts to adjust to be consistent with the model-input information.

Tucson Example

Boundary Conditions The **flow boundary conditions** applied to the model of Tucson are second and third type. In Figure 1.45 is illustrated the boundary-condition definition for one portion of the model-domain boundary. The definition of this boundary condition begins with the construction of a polygon that contains the portion of the domain outline defined in Section 1.2, for which a boundary condition is required. The definition of this segment is achieved by using the *Contour tool* when the *BC Flow L1* layer is active, as shown in Figure 1.45.

Once this area has been defined, it is necessary to specify the type and magnitude of the boundary condition. A double-click on the boundary-condition-defining polygon, or the last point used to define the polygon, brings up a *Contour Informa-*

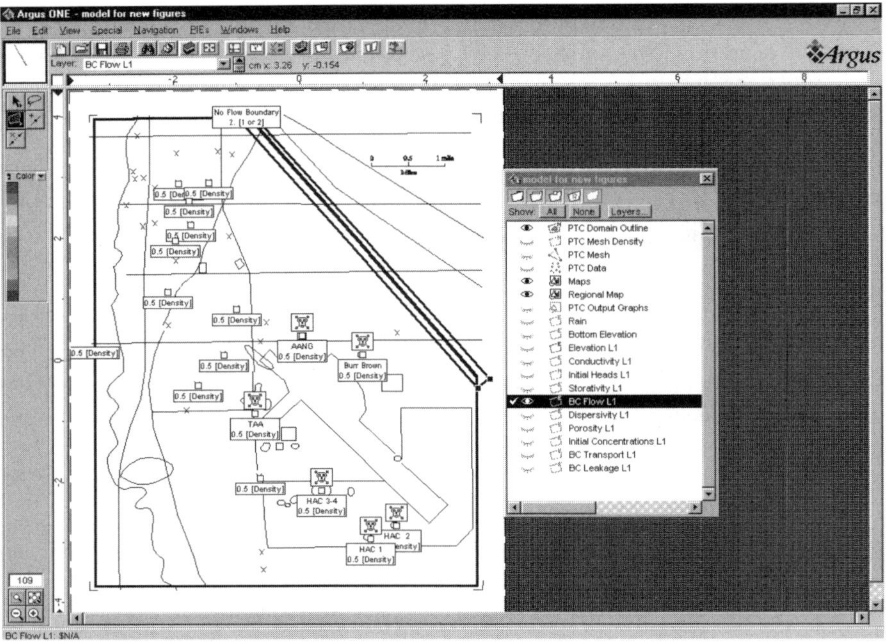

FIGURE 1.45. The boundary condition is defined within the polygonal area denoted by the squares at each vertex.

tion dialog box, indicated in Figure 1.46. One can specify a **first** (specified head) or **second** (specified head gradient) **type boundary**. This is emphasized by the value *1 or 2* that appears in the text box under *Units*. In the text box provided on the *BC Type L1* line, indicate the number of the boundary condition type, that is, either 1 or 2. In our example the number is 2 because we will specify a flux boundary con-

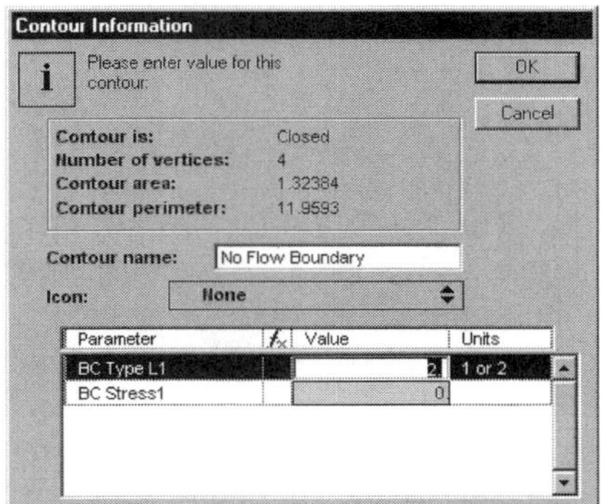

FIGURE 1.46. Contour-information dialog box used to define the type and value of the boundary-condition location defined by the portion of the domain boundary encompassed by the polytope with the black squares at each vertex.

dition along this segment of the model-domain boundary. Next, in the line identified by *BC Stress1*, place the magnitude of the specified gradient. Because the intent of the modeler is to have this segment of the boundary as a no-flow boundary, the value of the boundary condition will be 0.0. With these two numbers provided, one can now move to another portion of the boundary, or to a point on the interior.

To define a *point boundary condition*, such as that used to specify well discharge, we begin by copying the information we provided in Section 1.2 to the *BC Flow L1* layer. This is achieved by selecting the *PTC Domain Outline*, then selecting only the point contours where you plan to place boundary conditions for flow. Finally, select *Copy* from the edit menu.

Now activate the *BC Flow L1* layer and *Paste* the information obtained from the *PTC Domain Outline* layer. Double-click the placeholders you provided just now for each point information location. On a *Contour Information* dialog box similar to the one presented in Figure 1.46, fill in the text boxes, then click on *OK*. Keep in mind that if the point condition is a well (type 2) the units of discharge are volume per unit time. A series of point-boundary conditions are visible in Figure 1.47.

An alternative strategy is to follow the preceding protocol in reverse. That is, use the point tool to locate the point information locations on the *BC Flow L1* layer and then copy the information to the *PTC Domain Outline* layer.

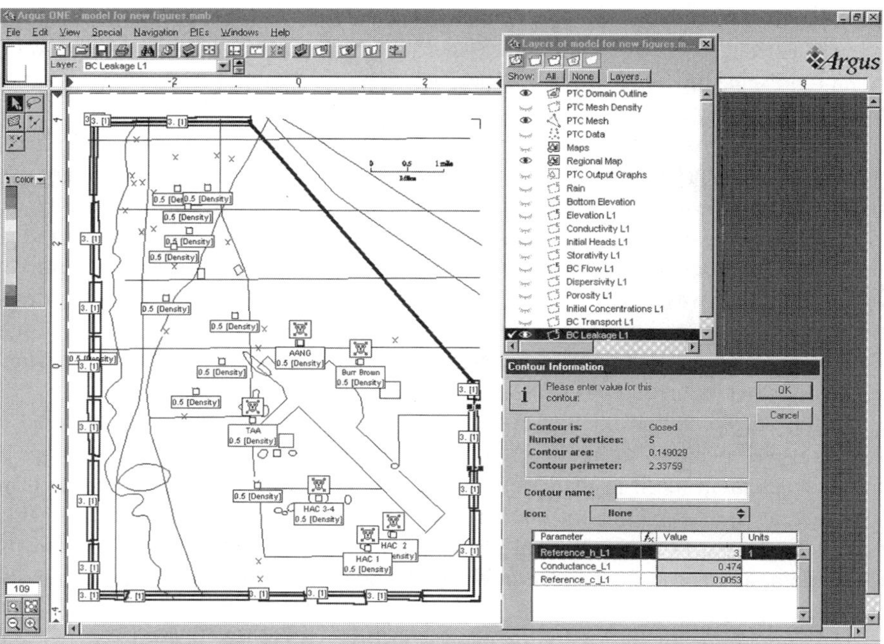

FIGURE 1.47. *The segment of the domain boundary defined by third-type or leakage boundary conditions. The contour information window is activated by double clicking a portion of the type-three domain definition polygon.*

In the case of the Tucson modeling project, the majority of the domain perimeter boundary was specified by the modeler as a **third-type boundary**. Third-type boundaries were described and discussed in Section 1.11.1. In that section we noted that a third-type boundary condition (Robbins) could be written as

$$K\frac{\partial h}{\partial n} = \kappa(h - h_0) \tag{1.71}$$

or

$$\frac{\kappa}{K}h - \frac{\partial h}{\partial n} = \frac{\kappa}{K}h_0 \tag{1.72}$$

which is a special case of the general third-type condition

$$\alpha h + \beta\frac{\partial h}{\partial n} = \gamma \tag{1.73}$$

where $\alpha = \kappa/K$, $\beta = -1$, and $\gamma = (\kappa/K)h_0$.

The interpretation of κ/K is usually one involving **vertical leakage** into the aquifer. The vertical flux from a surface water body is envisioned as occurring through a layer of sediment. The sediment layer is viewed as having a hydraulic conductivity and a thickness. Thus κ would have the physical interpretation of K_z/l, where l **is the thickness of the layer** and K_z **its hydraulic conductivity**. The value of h_0 represents the **head in the surface-water body**, a value which, while assumed to be known, may change over time.

However, as mentioned earlier, there is an entirely different way of looking at the third-type boundary condition. Suppose that one wants to extend the boundary of a model beyond the domain permitted by normal discretization. In other words, the analyst wants to make the model larger than is practically possible given computational limitations in order to accommodate the hydrological-boundary conditions observed in the field. One way to address such a case would be to imagine the boundary to be a long distance from the edge of the finite-element net. In this instance the horizontal flow into the model at the finite-element edge could be thought of in a mathematical sense as presented in equation 1.71.

In this interpretation, the reference head h_0 must be thought of as the head at some distant location, κ, as the effective hydraulic conductivity from the edge of the finite-element mesh to the location where h_0 is defined and l the distance from the edge of the finite-element mesh to the location h_0. In essence, by doing this, one is lumping all of the physical phenomena that occur between the edge of the finite-element mesh and the location where the boundary value h_0 is defined into the parameter κ.

In Figure 1.47 this interpretation of the third-type boundary is used. The various segments of the model boundary that are identified as third-type boundary conditions in order to extend the model domain are indicated in this figure. These segments normally have been introduced using the same procedure as described above for flow boundary conditions. The segments defined as third-type conditions are easily recognized by the number 3 that appears in the label for each segment.

To complete the specification of the third-type boundary, one must double-click each third-type boundary segment. This action activates the *Contour Information* dialog box indicated in Figure 1.47. Note that four pieces of information are requested for each boundary segment. The first is the label for the boundary segment *Contour name*, here left blank. The second request is for the reference head, which we have identified in equation 1.71 as h_0. The third request is for the conductance; this is the parameter identified as κ in equation 1.71. The fourth and final box is of no interest to us now, inasmuch as it is associated with the transport equation that we consider in Chapter 2. At this point we have specified all of the boundary conditions required for the flow equation.

Initial Conditions The initial conditions on the dependent variable head must be specified for each node in the model. Thus the appropriate protocol is to activate layer *Initial Heads L1* (Figure 1.48) and then use the approach described above for introducing information on the tops and bottoms of formations (see the Tucson example in Section 1.4).

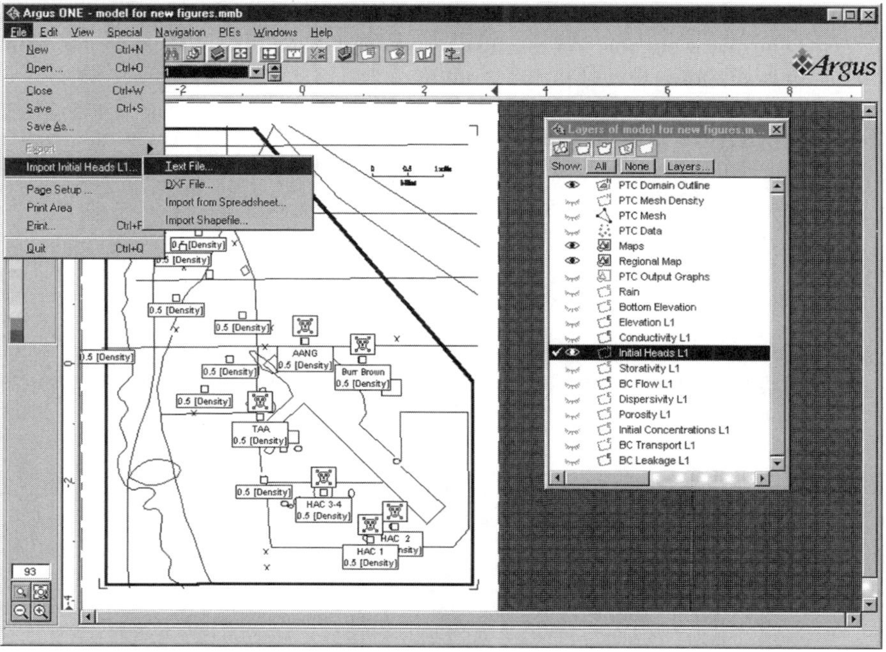

FIGURE 1.48. *Sequence of windows used to input initial conditions for hydraulic head as a text file.*

1.12 PARAMETERS

The flow equation contains two parameter fields, the **hydraulic conductivity K** and the **specific storage** S_s. When the simulator is areal two-dimensional, the **transmissivity T** and the **storage coefficient** S replace the hydraulic conductivity and the specific storage. In a water-table problem, porosity also plays a role.

In the event a finite-difference representation is used to approximate the flow equation, a value for each of these aquifer parameters must be provided at each node of the finite-difference model. Thus if there are N nodes in the finite-difference model, N values of each parameter must be input, one for each node. In the case of a body-centered finite-difference mesh, the parameter value can be thought of as representing information for the element in which a node resides.

In general, field information will not be available at each nodal location. It is therefore necessary to interpolate known information to form a surface that represents the required parameter and that will allow a value to be input at each nodal location. This can either be done by hand or by a computer-based interpolation program. **Kriging** is often the algorithmic engine used to do computer-generated interpolation. However, as is evident from our Tucson example, Argus ONE has built-in interpolation capability.

In the case of a finite-element mesh, one has the option not only of specifying parameter values at the nodes but it is also possible to provide a constant value over an element. In the finite-element approximation to the flow equation discussed earlier (see equation 1.66) we assumed that the parameter values were constant over each element. Thus we were able to pass the constant parameter through the integral sign. The result is simplification of the integration.

If one elects to specify parameter **information by node**, the following strategy can be used to perform the required elementwise integrations. Represent the nodally defined parameter using the basis functions employed to represent the unknown state function (or an alternative set as appropriate), visualize

$$\mathbf{K}(\mathbf{x}) = \sum_{k=1}^{2} \mathbf{K}_k \cdot \phi(\mathbf{x})_k \tag{1.74}$$

where the number of nodes, in this case two, is suitable for our linear one-dimensional finite-element example. Using this approach, we obtain for a given finite element (see equation 1.66),

$$-\int_e \left[\sum_{k=1}^{2} \mathbf{K}_k \cdot \phi(x)_k \frac{d}{dx} \sum_{j=1}^{J} h_{n+1,j}\phi(x)_j \frac{d}{dx}\phi(x)_i \right] d\mathbf{\Omega}_e$$
$$-\int_e \left[\sum_{i=1}^{2} S_k\phi(x)_k \sum_{j=1}^{J} \frac{h_{n+1,j} - h_{n,j}}{\Delta t}\phi(x)_j + Q \right] \phi(x)_i \, d\mathbf{\Omega}_e \tag{1.75}$$

The use of the nodewise definition of parameters is particularly important when using nonlinear coefficients such as the thickness $l(h)$ indicated in the water-table equation 1.35. When this situation arises it is convenient that the unknown head values are obtained at the nodes, since this clearly facilitates, and encourages, obtaining the nonlinear coefficients at the nodes. Once the nodal values of the nonlinear coefficients are available, they can easily be accommodated using the formulation provided in equation 1.75.

1.13 FRACTURED AND CAVERNOUS MEDIA

Equation 1.7, the groundwater-flow equation, is applicable only when the aquifer can be considered, from a mathematical physics perspective, as a porous medium. In essence, this means that the various material properties descriptive of the aquifer, and consequently the resulting hydraulic head solution, must be representable by a smooth function. In other words, it is assumed that there are no discontinuities in the model parameters. Most porous media satisfy this assumption, their properties varying areally but not in a discontinuous fashion. An exception might be identified at an aquifer-aquitard boundary, but even here the gradation is continuous and limited in its areal extent.

Fractured media do not satisfy the assumptions stated above. Fractures have a profound impact on the movement of groundwater through aquifers, especially aquifers with relatively low primary hydraulic conductivity (we discussed primary and secondary hydraulic conductivity in Section 1.4.2). In general, the fractures as a medium, have a large hydraulic conductivity in comparison with that of the host rock or blocks. On the other hand, the host rock can have a large storage capacity. In other words, fractures are rapid conduits for groundwater, but do not, in and of themselves, hold much water. On the other hand, the intervening blocks may have a relatively low hydraulic conductivity, but hold a large amount of water.

Given the existence of these quite different properties within the same rock mass, the assumption of smoothly varying material properties is generally not satisfied. Consider, for example, the abrupt change in hydraulic conductivity that one would encounter moving across the boundary between a block and a fracture. To accommodate the schizophrenic nature of the fractured media, one can proceed in either of two quite different ways.

One approach is to assume that two media coexist at the same mathematical point. Although intuitively odd, this is an acceptable mathematical concept provided that one is always thinking of the properties as existing as an average over a defined, although arbitrary volume. In other words, in a sufficiently large volume both fractures and blocks may coexist, and therefore there exists an average hydraulic conductivity for both the fractures and the blocks. Moreover, one could identify these properties with a point in the averaging volume, for example the centroid. Now there are two equations similar to equation 1.7, one for the fractures and the other for the blocks. Of course, these systems are coupled. Water and solute move from the fractures to the porous blocks, and vice versa. Thus there must be a coupling term in each equation. This term will look like a source (or sink) and will describe the movement between the fractured-rock system and the porous-block system. Each node in the model will have two head values and two concentration values, one associated with each system. The conceptualization is often called a *dual-porosity approach*.

The second way to approach the modeling of fractured systems assumes that the geometry of the fractures is known, as are its hydrodynamic characteristics. For example, one could replace hydraulic conductivity with fracture surface roughness and describe the flow in the fracture using a different set of governing equations. The fracture wall would be a boundary condition on the fracture system and the block. Each fracture would be defined uniquely, and its properties would not be considered as averages as they were in the dual porosity system defined above. The walls of the fractures are treated as the locations of boundary conditions for both the fracture and block equations. The boundary conditions define the coupling. Of course, the crunch in this method is knowing the geometry of the fractures, a significant challenge in the field. We call this second strategy the *discrete-fracture approach*.

Important to us is the fact that the flow and transport models described in this book are only applicable when the smoothness conditions described above are met. In other words, if equation 1.7 is to be used, the fractures must be sufficiently ubiquitous so that they appear, on the average, uniform. For this assumption to hold may require very large scale modeling. It would not be appropriate to use a model based on equation 1.7 for an aquifer system that had only a few dominant fractures. The fractures must be sufficiently frequent that the spatially moving average of parameters describing them is smooth in a mathematical sense.

One can use the discrete fracture approach using the formulation of equation 1.7 by taking advantage of the features inherent in both the finite-difference and finite-element methods of solution. In the case of finite difference, fractures can be represented by lines or planes of elements that have the properties of the fractures. In the case of finite-element methods, there are additional options. It is possible to define one- or two-dimensional elements that exhibit the parametric behavior of fractures. These elements can then be coupled together formally with the three-dimensional prism elements to represent the fracture behavior.

Cavernous rock units act for all intents and purposes as fractured media, although the size of the caverns may exacerbate the impact they have on the aquifer behavior. In essence, all of the preceding discussion on fractured media is transferable to cavernous media. In general, if equation 1.7 is to be used as the basis for a model of a cavernous medium, only models representing very large areas will satisfy the fundamental mathematical assumptions inherent in this equation.

Tucson Example

Hydraulic Conductivity In Section 1.4.5 we found that the hydrogeological units in the Tucson case-study area changed their characteristics within the area of the proposed model. More specifically, we find the following reported (Black and Veatch [10]):

> There is a general increase in coarse-grained sediments from the east toward the west, particularly in the interval that comprises the upper aquifer. This pattern of coarse-

> and fine-grained sediments is expected to have an important influence on the rate and direction of groundwater movement. Groundwater is expected to move more rapidly through the section containing the larger fraction of coarse-grained deposits, which occur on the western side of the property.
>
> Hydraulic gradients observed from groundwater levels can provide information on relative variations in the water transmitting properties of an aquifer. Groundwater levels in the TAA have reached a near steady-state condition in the upper aquifer as indicated by groundwater level measurements taken by the ADWR over the last several years. Plate 4 [see our Figure 1.49] shows a contour map of groundwater levels in the upper aquifer for 1984 as prepared by Mock et al. [6]. This map indicates variability in hydraulic gradients, which are expected to be related to variations in the transmissivity of the aquifer, which is a function of the hydraulic conductivity and saturated thickness of the aquifer. In those areas where the contours are close together, the transmissivity is expected to be lower than in those areas where the contours are more widely separated.
>
> Plate 4 indicates that several zones of varying transmissivity exist in the TAA. In general, the transmissivity of the upper aquifer appears to increase to the west. A zone of less transmissive material appears to occur in a somewhat north–south pattern almost parallel to the Old Nogales Highway. Groundwater level contours in the eastern portion of the upper aquifer appear relatively uniform; however, groundwater level measurements are more sparse in that area.
>
> A number of aquifer tests have been conducted in the TAA as a part of the TAA RI and Air Force's IRP. Table 4-1[25] presents the results of the tests as reported by Mock et al. [6]. . . . These test results indicate a range in estimated hydraulic conductivity values of over three orders of magnitude. The lowest value of 3 gallons per day per square foot (gpd/ft^2, 0.4 ft/d) is reported for Monitoring Well WR-55B; and the highest value of 2,000 gpd/ft^2 (270 ft/d) is reported for Monitoring Wells M-8 and TAS-9. This range in hydraulic conductivity values is expected given the variability observed in the geologic materials, which is illustrated in the geologic cross sections described previously. The geometric mean value from the 52 values is 332 gpd/ft^2 (44 ft/d).

Based on the work of Mock et al. [6] the modeler, Dr. Spiliotopoulos, divided the aquifer area into zones, each with a specified and constant hydraulic conductivity as illustrated in Figure 1.52. The construction of this diagram and the mechanism for introducing the porosity and storage coefficient discussed in the following two sections using Argus ONE are discussed in Section 1.13.

Porosity The value of the **porosity**, or more specifically the effective porosity, of the system[26] was selected by Mock et al. [6] based on the following:

> The effective porosity of a clean sand and gravel aquifer is practically identical to its volumetric porosity because dead-end pore space[27] and adhesion[28] are negligible

[25] The table is not included in this book; see the original reference.

[26] *Effective porosity* is the volume of the void space in the aquifer through which there is fluid movement.

[27] *Dead-end pore space* refers to that portion of the void space in an aquifer where there is no fluid movement. It is often envisioned as an isolated pore which has only one opening to an adjacent pore.

[28] *Adhesion*, in this context, refers to the effect of molecular forces of attraction between the fluid in the pores and the porous solid (i.e., the grain surfaces).

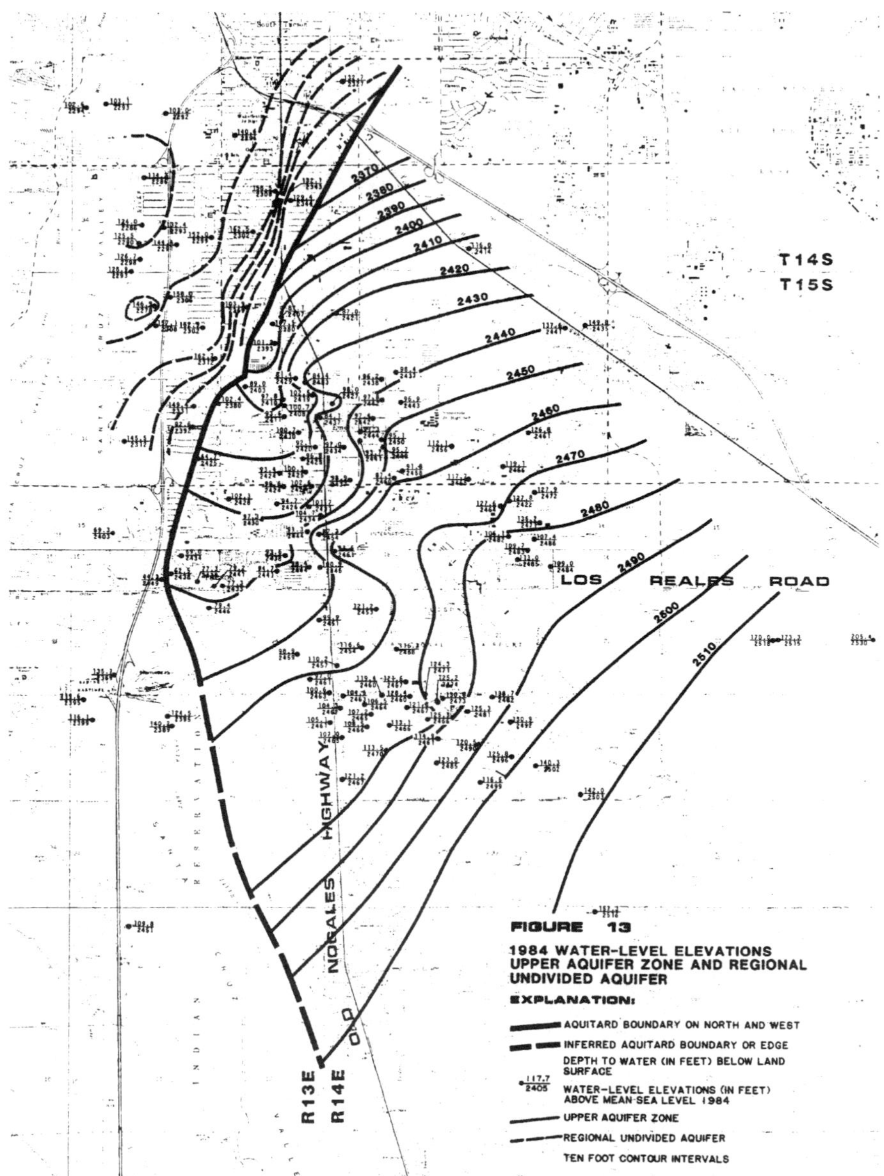

FIGURE 1.49. *Water-level elevations observed in 1984 for the upper aquifer and regional undivided aquifer [6].*

(Freeze and Cherry [11]). However, in clays or heavily cemented sediments, effective porosity can depart significantly from the volumetric porosity. The specific yield (the unconfined storage coefficient or drainable porosity if the sediments were to be drained) can provide an estimate of the lower limit to effective porosity.

Specific Yield < Effective Porosity < Volumetric Porosity

For the TAA, upper aquifer zone materials range from clays and sandy-clays to sands and gravels. The estimated specific yield of sediments in the upper aquifer zone range from 0.05 to 0.25 and average approximately 0.15.

Storage Coefficient In general, the information on **storage coefficients** is relatively scarce. Such is the case in the Tucson model area. In the model prepared by Mock et al. [6], we read the following:

> The model study of Travers and Mock [12] analyzed the distribution of storage coefficient values throughout the Upper Santa Cruz basin. Comparisons of sediment descriptions to published estimates completed for that study indicated that the storage coefficient ranges from 0.05 to 0.25 in the upper 200 feet of sediments in the TAA and averages approximately 0.15. The model chosen for this study only accepts a uniform storage coefficient.[29] Time allocated for this modeling effort did not allow for reprogramming of the model code to include distributed storage coefficient values. Any appropriate value for storage coefficient could have been selected, but 0.15[30] has been used in previous model studies in the area (Hargis and Montgomery [13]; Anderson [14]).

Parameter Input As in the case of the **formation elevation information** discussed in Section 1.4.5, parameter information can be passed to *PTC* via the Argus ONE environment in several ways. One approach is to provide a constant value over the entire region of interest for a given geological layer. To achieve this goal one first selects *View* from the menu bar and then, from the options that are made available in the view menu, *Layers*. By selecting *Conductivity L1*, the *Layer Parameters* dialog box is opened. The result is shown in Figure 1.50. Notice that in the *Layer Parameters* window there are three variables specified: *xConductivity L1*, *yConductivity L1*, and *zConductivity L1*. The specification of these three components of the hydraulic conductivity allows one to accommodate anisotropy, provided that the off-diagonal components are zero. The off-diagonal components are, indeed, zero when the coordinate axes are collinear with the principal directions of the hydraulic conductivity tensor. As can be seen from Figure 1.50, the default values for condl1y and condl1z are condl1x; this set of values represents the special case of an isotropic aquifer.

To specify a value for *xConductivity L1*, and so on, one clicks on the f_x button. This gives rise to the window presented in Figure 1.51. Several of the Argus ONE facilities can now be used to specify the function *xConductivity L1*. In this example,

[29]Note that this is not the case for PTC.

[30]The value of 0.15 is very large for a storage coefficient and probably reflects the influence of dewatering (drainage) of the upper portions of the aquifer.

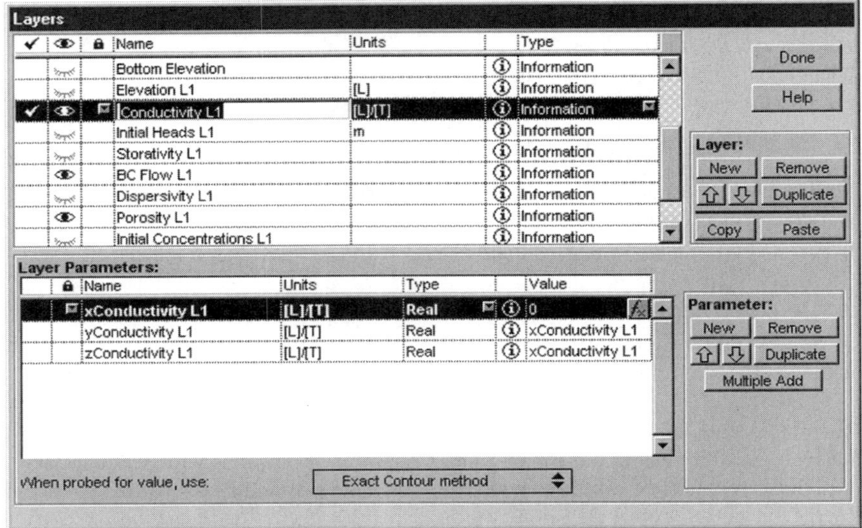

FIGURE 1.50. *The selection of View followed by Layers provides the indicated window. When the Conductivity L1 option is chosen the Layer Parameters: dialog box found as the lower window illustrated in this figure is revealed.*

we have simply entered an integer. Other options are described in the Argus ONE manual.

When a **distributed hydraulic conductivity field** is required, which is normally the case, an effective strategy is to import a picture of a hydraulic-conductivity field as presented in a published document as a first estimate. Using this picture as a base map, contours can be drawn using Argus ONE in a manner similar to that used for the formation top and formation bottom contours in the Tucson example in Section 1.4.5.

Using the contour tool the hydraulic conductivity zones illustrated in Figure 1.52 can be created. The specific value to be assigned to the area within a specified domain is established by double-clicking the contour and assigning the appropriate value. In this example, the area within a domain is assigned the value identified with the domain outline. Later, after the finite-element mesh is created, one can verify that this is indeed the case by double-clicking on a node within the domain of interest and viewing the information associated with that node. The creation of the finite-element mesh is discussed in Section 1.15.

Another convenient way to input hydraulic conductivity information is using the *Point* tool. To do this, one identifies each location where a hydraulic conductivity value is known and when prompted provides the specific value. The interpolation algorithms can then be used to create the field. There are several variants on this theme, which are described in the Argus ONE manual.

The other two parameters of interest, the specific storage (or storage coefficient in two-dimensional areal problems) and porosity, can be introduced into the model in a similar manner by activating the *Storativity L1* and *Porosity L1* options, respectively.

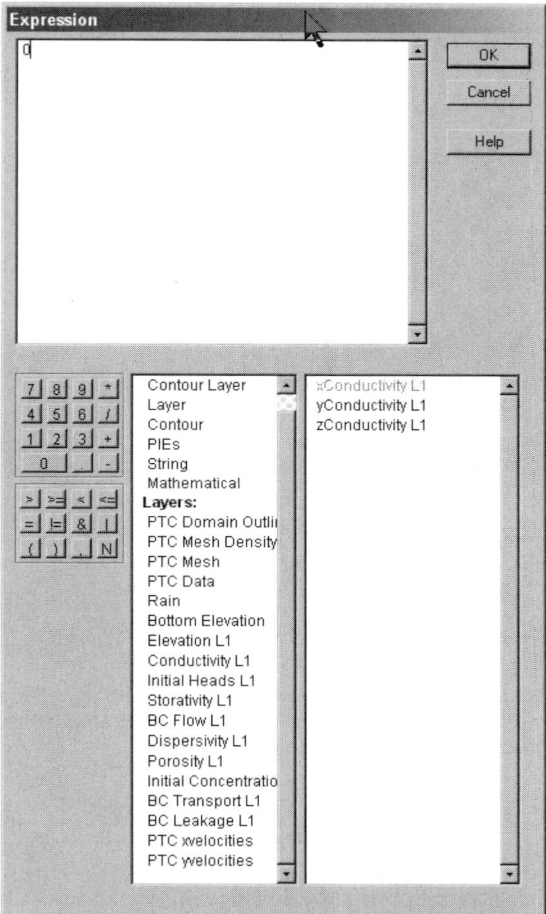

FIGURE 1.51. *By clicking the f_X window one obtains the window found in this figure. One can now specify the value of condl1x using several of the Argus ONE features.*

1.14 MODEL STRESSES

Model stresses can generally be cataloged as *pointwise* and *areally distributed*. From a theoretical point of view, pointwise stresses are handled quite differently in finite-difference methods than in finite-element methods. However, from a practical point of view, both methods boil down to assigning the stress to a node. In the case of finite-difference methods, the basic concept is simply to assign to a node the volumetric flux associated with that node. For example, if the stress was a pumping well discharging at a rate of 1000 gallons per minute, one would simply replace Q in equation 1.7 by the value 1000 divided by $\Delta x \cdot \Delta y \cdot \Delta z$. To see why one divides by the volume $\Delta x \cdot \Delta y \cdot \Delta z$, it is helpful to remember that the time derivative in the

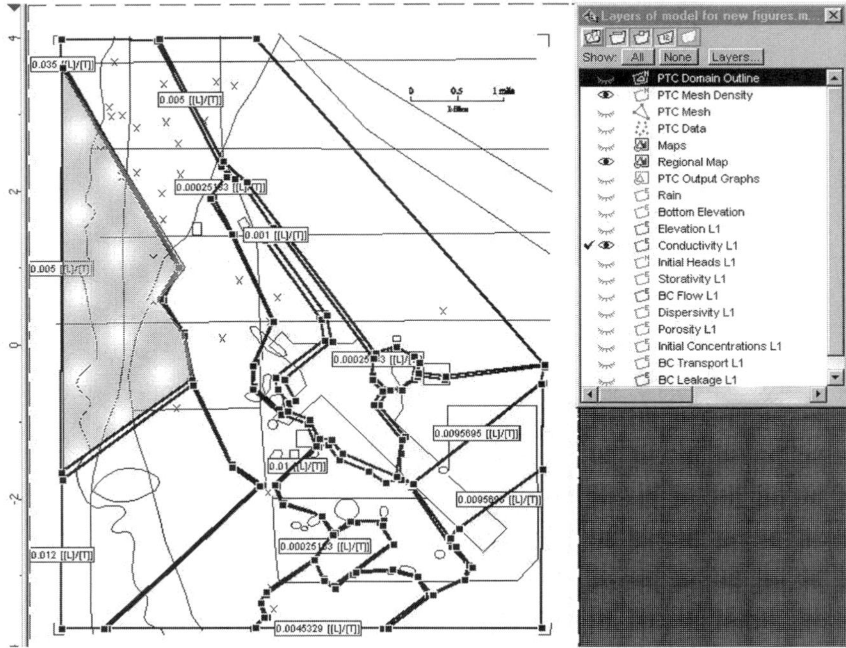

FIGURE 1.52. Hydraulic-conductivity domain outlines. In this example all the values within the domain outlines are of the same hydraulic conductivity. The shaded area, for example, represents hydraulic conductivity values of 0.005 m/day.

flow equation has units of $1/t$. Thus, to accommodate a volumetric flux defined in terms of volume per unit time, the appropriate scaling factor is $\Delta x \cdot \Delta y \cdot \Delta z$. In the case of a two-dimensional problem, the divisor would be the area associated with the finite-difference node, $\Delta x \cdot \Delta y$, for example.

In the case of finite elements, one must return to the original equation to arrive at an appropriate formulation. Consider, for example, equation 1.57:

$$\int_{\Omega} \left(\nabla \cdot \mathbf{K} \cdot \nabla \hat{h} - S\frac{\partial \hat{h}}{\partial t} - Q \right) \cdot \phi(\mathbf{x})_i \, d\Omega = 0, \qquad i = 1 \dots I$$

The recharge term is defined by Q. In the case of a *point source*, the appropriate expression is $Q\delta(\mathbf{x} - \mathbf{x}_i)$, which says that the volumetric flux Q is specified at the point $\mathbf{x}_i$. Substituting this definition into the integral form we obtain for the source

$$\int_{\Omega} Q\delta(\mathbf{x} - \mathbf{x}_i) \cdot \phi(\mathbf{x})_i \, d\Omega \tag{1.76}$$

which, by definition of the Dirac delta function becomes $Q(\mathbf{x}_i)$, the discharge defined at the node located at $\mathbf{x}_i$. Thus, in the case of the finite-element method, the accommodation of a source involves simply the specification of the source strength. In other words, one specifies the volumetric discharge for a well located at a specified node.

In the case of a *spatially distributed source*, the situation is quite different. One must distribute the source over the entire element. In the case of finite-difference methods, one assigns to each node a value equal to the flux per unit area. Once again the physical dimensions of the assigned values must be $1/t$ in a three-dimensional model.

In the case of a finite-element model there are two choices. To see these choices let us examine the source term in the groundwater flow model, visualize

$$\int_e Q\phi(\mathbf{x})_i \, d\boldsymbol{\Omega}_e$$

If we assume the Q is constant over the element e, this integral becomes

$$Q \int_e \phi(\mathbf{x})_i \, d\boldsymbol{\Omega}_e$$

which can be readily evaluated provided Q is known. If, however, the flux varies over the element as might be the case when leakage due to leaking pipes or sewers is encountered, it is necessary to assume a functional form for Q, that is

$$Q(\mathbf{x}) = \sum_{k=1}^{E} Q_k \phi(\mathbf{x})_k$$

where N_E is the number of nodes in element e. The appropriate integral now becomes

$$\sum_{k=1}^{N_E} \int_e Q_k \phi(\mathbf{x})_k \phi(\mathbf{x})_i \, d\boldsymbol{\Omega}_e \tag{1.77}$$

In this case, the values of the flux are nodal values and the integration involves the product of basis functions. This is the same form of integral that occurs in the case of the time derivative.

The nodal value Q_k can be a function of the unknown head, such as is the case when one is working with a Robbins-type boundary condition. In this case, the resulting additional unknowns appearing in equation 1.77 will be transferred to the right-hand side of the equation and the integrals will become part of the coefficient matrix.

1.14.1 Well Discharge or Recharge

The hydrological stresses required in the Tucson model would consist of well pumpage and reinjection, and rainfall. Well pumpage is treated in the form of a boundary condition and is discussed in Section 1.11.1. As mentioned earlier, the units applicable to this input are volume per unit time, the specific units being defined by those used for the scaling and parameters. **Well discharge is identified by a negative sign in front of the value**.

1.14.2 Rainfall

Rainfall (actually net infiltration from any areally distributed source) is accommodated as a spatially-variable input similar to the parameters and thickness input information discussed earlier. The *Rain* layer contains the net-infiltration information for layer one. In a multiple-layer model, only this layer would have rainfall input. To introduce rainfall make the *Rain* layer active. If a **global** (areally constant) value of rainfall is to be used, select *View* from the menu bar and then *Layers...* from the view menu. Alternatively, you can click on the *Layers Dialog* tool, the tenth icon from the right on the toolbar. This brings up the *Layers* dialog box. By clicking on the f_x option, a workspace opens into which the global value of *Rain Stress1* can be placed in units of length over time.

If a spatially variable rainfall is required, one first opens the *Rain* layer and then uses the *Contour* tool to define the areas wherein a constant value of rainfall is to be defined. Upon completion of the contour with a double-click, a *Contour Information* dialog box opens that permits you to enter the rainfall value *Rain Stress1* by clicking on the box under the word *Value*. This process is repeated for each contoured region. When the last contour is defined and the values provided, the rainfall input is complete.

1.14.3 Multiple Stress Periods

To this point we have tacitly assumed that all the information entered using the GMA approach was time independent. In other words, the information to be used in the model would not change over the period of analysis to be considered. Such an assumption is normally unwarranted. Thus the ability to consider several periods of time in which each can take on different parametric values is important. To address this issue, we must return to the *PTC Configuration* window, which is accessed by selecting *PIEs* from the toolbar in the main window. From the *PIEs* menu, select *Edit Project Info...* On the *PTC Configuration* window select the *Stresses* tab. The window similar to that appearing in Figure 1.53 results.

The difference between the window obtained using this sequence of operations and that presented in Figure 1.53 lies in the information found in the workspace located below the *Stresses* tab. Now, rather than having only one line of values in this workspace, we now have two. The first line is identical to that obtained earlier. The second represents information identified with the second stress period. In the column beneath the heading *Stress*, two stress periods are identified, stress period 1 and stress period 2. In the column labeled *Flow*, there appear two ones. This indicates that the *Do flow* check box has been activated during both stress periods. Thus all of the information pertinent to flow, for example pumping rates, will have two sets of

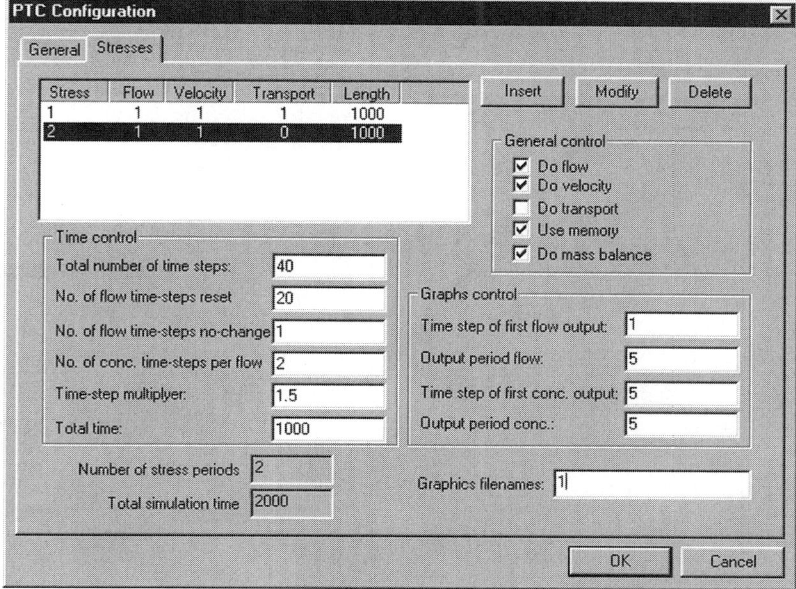

FIGURE 1.53. *PTC Configuration window with the Stresses option selected.*

values, one for each period. Similarly, a velocity calculation has been requested for each stress period. However, in the *Transport* column a zero appears in the second stress period. This is consistent with the information appearing in the *General control* check boxes. The check box for transport has been turned off.

The final column presents the length of each of the stress periods. In this case they are both 1000 time units long. In the lower left-hand corner of the window the *Number of stress periods* and the *Total simulation time* are conveniently tabulated.

As noted above, relevant information must be provided for each stress period. This includes the values appearing in the *Time control* and *Graphs control* text boxes. Each stress period can potentially have different information in these text boxes. We discuss the contents of these text boxes in Sections 1.16 and 1.17.

To this point we identified model input as associated with stress period 1. Now we realize that this is not always going to be the case. As an example, we provide in Figure 1.54 the specifications for two pumping-well flow rates, one for each of the two stress periods.

Tucson Example

In general, and at Tucson in particular, there are two sources of infiltration, natural and human-induced. Human-induced can be deliberate, such as through irrigation, or accidental, as would be the case were a water-carrying pipeline to leak. The following

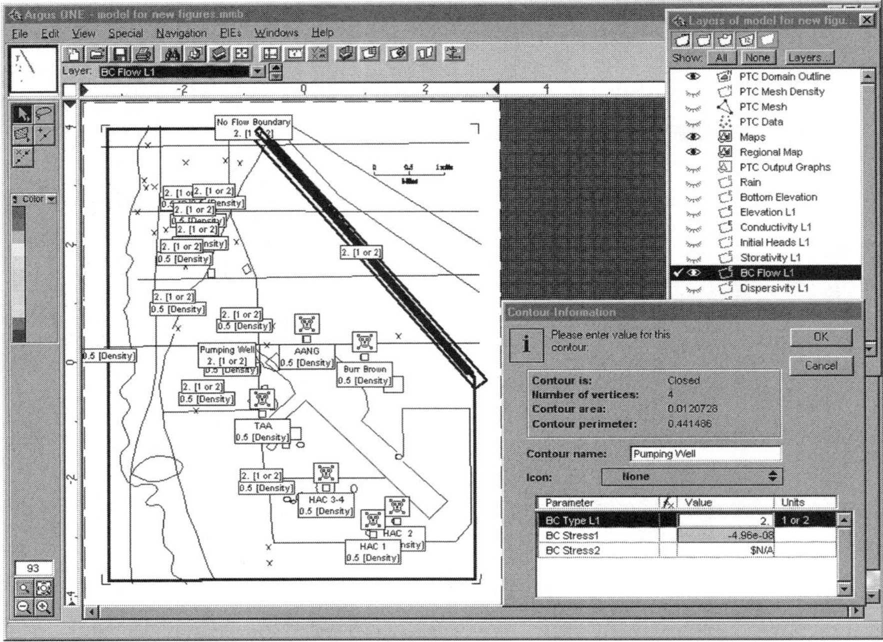

FIGURE 1.54. *Input of multiple (2) well pumping stresses. Note the Contour Information dialog box opens when the well location identified by the label Pumping Well is double clicked.*

discussion by Mock et al. [6] describes their view on the treatment of recharge in the Tucson model area:

> Recharge input values were classified as natural or incidental. Natural recharge included recharge from seasonal stream flow events and mountain front recharge. Natural recharge estimates were taken from Travers and Mock [12]. Tributary washes above the upper aquifer zone carry small volumes of natural surface flow. Amounts of recharge from them could not be quantified but were thought to be insignificant. Incidental recharge in this area includes recharge occurring through application of irrigation water to agricultural lands.... Recharge was thought to be negligible in the upper aquifer zone because generally recognized recharge areas at the surface were thought to be very small or absent.
>
> Source investigation work completed by Rampe [4] presented information on incidental recharge from historic wastewater disposal at the HAC facility.

1.15 FINITE-ELEMENT MESH

From the information provided in Sections 1.10 and 1.11 we have a sense as to the important role that **mesh configuration** plays in the accuracy obtained with a numerical simulator. We realize that a **mesh** must be more refined in those areas where we anticipate that the potential surface will have the greatest curvature. Given a constant hydraulic conductivity, this will also be the area where one can anticipate the highest groundwater velocities. To illustrate the protocol to be followed in defining a mesh for *PTC* using Argus ONE, let us now return to the Tucson example.

Tucson Example

Regional Mesh If one employs the Argus ONE interface, and the number of physical dimensions are specified, discretization of the model is determined largely by the Argus ONE software. More specifically, having defined the number of **vertical layers** in the model, the finite-element mesh created by Argus ONE minimizes the numerical **discretization error** in the simulation. One does, however, have control over one of the variables defining the characteristics of the calculated mesh. One can, and indeed must, specify a **density** parameter that will define the global mesh spacing and therefore the **numerical accuracy** of the model.

The density parameter is defined in terms of the screen coordinates. If the horizontal length of the model is 10 units, a density specification of 1 will generate elements that "in some sense" will have an average length of 1. On the other hand, if the overall length is 20 units and the density is chosen to be 2, the average length of an element is 2. The existence of point-value specifications will modify this overall assessment since the point values will result in smaller elements in the neighborhood of the point value.

From the analyst's point of view, the important issue is the overall density value. A physically complex problem will require more elements to obtain an accurate simulation. As a rule of thumb, one should use elements that are on the order of hundreds of feet in terms of field units, unless the groundwater-flow pattern is very complex. A density that is on the order of tens of feet may be required when a complex flow pattern exists or is anticipated. Keep in mind that the information input into the model is in screen units and an appropriate transformation from field to screen units is needed to arrive at a suitable finite-element density.

To provide the density information needed by Argus ONE, one first makes the domain outline layer active. Next, double-click on the domain outline. The window shown in Figure 1.55 appears. In the text box in the *Contour Information* dialog box, place the density. Here the density is given as 0.5 screen unit. As demonstrated below, this value assigned to each boundary, including point boundaries, gives an acceptable mesh. This figure also shows the density values assigned to each of the point information contours, that is, the locations identified with the wells and the point sources for contaminants.

Now select the *PTC Mesh* layer and make it active. Select the *Magic Wand*, the highlighted icon in the mesh tool kit, and drag it to a point inside the domain-outlined area. A finite-element mesh consistent with your specifications is generated (Figure 1.56). If the mesh does not seem appropriate, that is there are either too few or too many elements, one can once again make the *Domain Outline* layer active, click on the appropriate domain outline, and when the appropriate window opens,

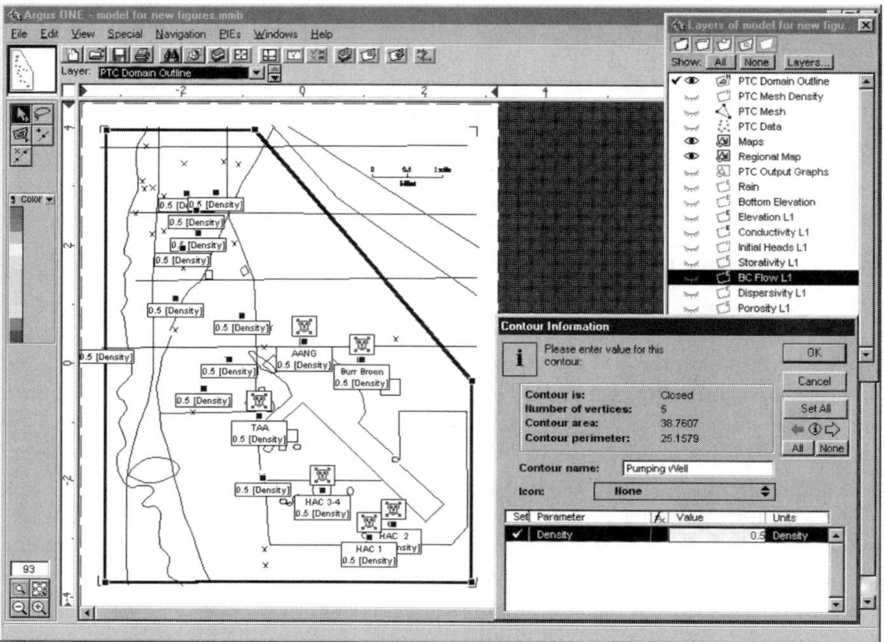

FIGURE 1.55. Window introduced by clicking on the domain outline (or point domain) locations to introduce the density values. The appropriate value is placed in the text box in the Contour Information dialog box.

change the density. Since Argus ONE creates and numbers the nodes and elements in a mesh very rapidly, it is possible, and indeed advisable, to take time at this stage to obtain the best element configuration consistent with the needs of the analysis and the available computational capability. A decision regarding the suitability of a given mesh will be more informed if viewed within the context presented in Sections 1.10 and 1.11.

It is evident from examination of Figure 1.56 that the size (and therefore the density) of the element array changes over the model area. However, on average, it would appear that the length of an element side is about 0.5 screen unit (the units shown at the top of the workspace), as we requested in the density specification. Most evident is the **impact that singular points** have on the element array. Inasmuch as the groundwater surface can be expected to change dramatically in the neighborhood of wells, the finite-element mesh must be finer in order to capture this curvature accurately. In areas where the groundwater surface is expected to be relatively flat, a coarser mesh is generated. Notice also that a singular point is defined by a node. The rationale behind locating a node at each singular point is that all boundary conditions, including well discharge and recharge, must be defined at a node.

Refining the Mesh Let us assume that having observed the mesh generated by Argus ONE, there is a perceived need to refine the mesh in a selected area. To do

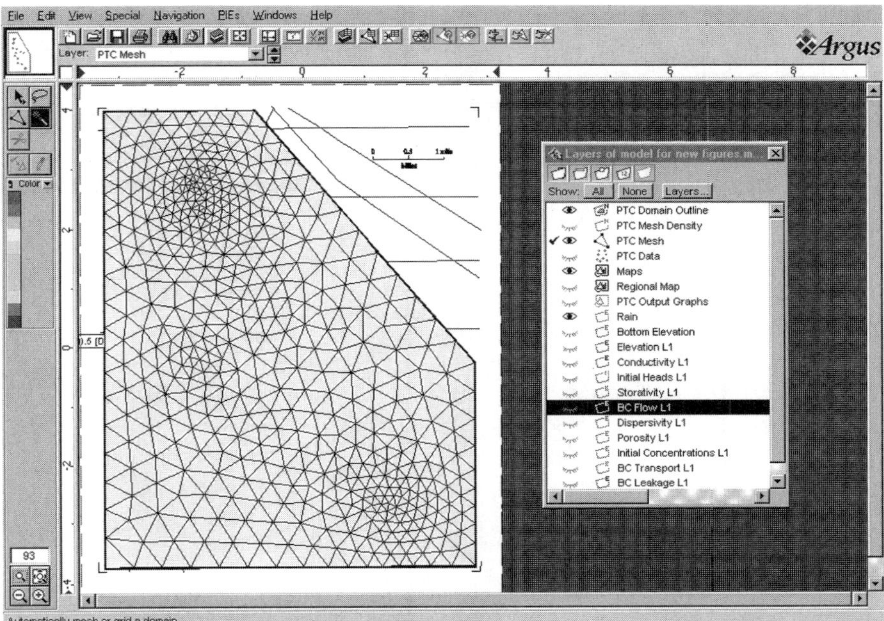

FIGURE 1.56. *Finite-element mesh generated by making the PTC Mesh layer active. Note that the mesh is more dense (exhibits more elements) in the neighborhood of singularities such as wells. The resulting denser mesh improves the accuracy of the model in these locations.*

this, one first selects the *Mesh layer* to reveal the existing mesh configuration. The *PTC Domain Outline* layer is now made active. Using the contour tool, the area to be refined is defined. Next in the window of the *Contour Information* dialog box place a smaller number in the text box defined by row *PTC Mesh Density* and column *Value*. Assume that a value of 0.1 is selected, a value that is one-fifth of the global value of 0.5.

To create the modified mesh, first make the *Mesh layer* active. Select the *Magic Wand* from the tool kit available at the upper left-hand corner of the screen. Drag the wand to the area defined by the domain outline to be refined, and drop it. When the dialog box opens, select the option to delete all. The result will be a completely-reformulated mesh that respects the requested refinement in the designated area. The result of the refinement is shown in Figure 1.57.

Whenever a mesh is modified, the numbering of the nodes changes. The resulting numbering will not, in general, be optimal from the point of view of minimizing bandwidth and therefore computational effort (see Section 1.16.2). To return to an optimal numbering of the nodes one activates the *PTC Mesh* layer, then selects the *Special* menu and menu item *Renumber*. In the *Renumber* dialog box check *Optimize Bandwidth*. Note that the *Optimize Bandwidth* option is highlighted only after the domain has been remeshed. Failure to optimize bandwidth often leads to failure of the matrix solver.

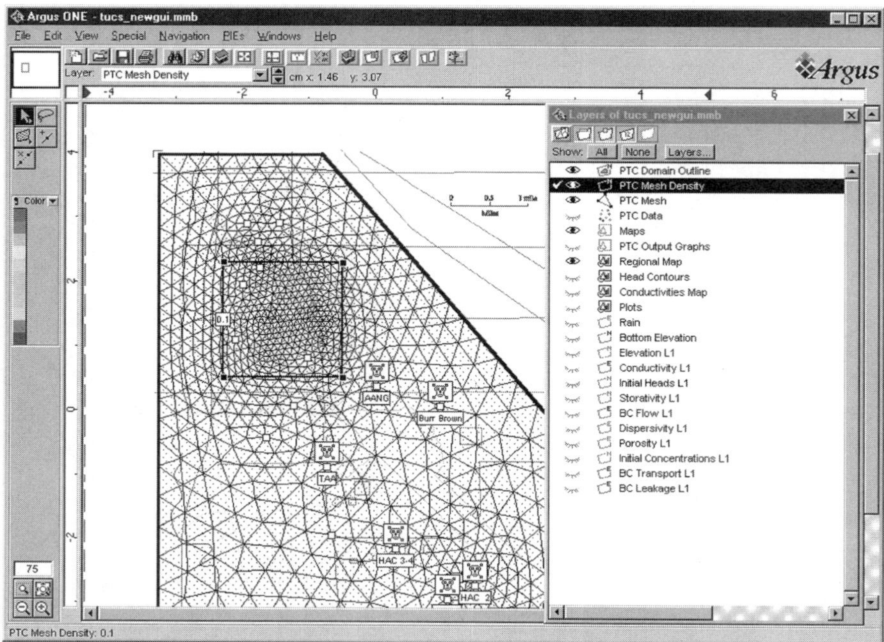

FIGURE 1.57. *The outcome of refining the mesh in the area designated by the square in the upper portion of the domain. This refined area should have elements that, on the average, have a length of 0.1 as compared to the remainder of the mesh that has elements that average 0.5 length units (screen units).*

1.16 SIMULATION

In this section we first discuss the matrix equations generated when the finite-difference or finite-element method of approximation is applied to the flow equation. Subsequently, we explain the method used to solve the resulting equations.

1.16.1 Solution Algorithm

We begin by restating the groundwater flow equation:

$$\nabla \cdot \mathbf{K} \cdot \nabla h = S_s \frac{\partial h}{\partial t} + Q$$

and rewrite this equation as

$$(L_x + L_y)h + L_z h = \frac{\partial h}{\partial t} + \frac{Q}{S_s} \tag{1.78}$$

where, for example,

$$L_x h \equiv \frac{1}{S_s}\frac{\partial}{\partial x}\left(K_{xx}\frac{\partial h}{\partial x}\right)$$

Let us now use a finite-difference approximation in equation 1.78 to obtain[31]

$$(L_x + L_y)h^{n+\theta} + L_z h^{n+\theta} = \frac{h^{n+1} - h^n}{\Delta t} + \frac{Q^{n+\theta}}{S_s} \tag{1.79}$$

where $h^{n+\theta} = \theta h^{n+1} + (1-\theta)\, h^n$ with $0 \le \theta \le 1$. For convenience we write equation 1.79 as

$$L_{xy} h^{n+\theta} + L_z h^{n+\theta} = \frac{h^{n+1} - h^n}{\Delta t} + \frac{Q^{n+\theta}}{S_s}$$

where $L_{xy}h^{n+\theta} \equiv (L_x + L_y)h^{n+\theta}$. One can rearrange equation 1.79 to obtain

$$[I - \theta(\Delta t)(L_{xy} + L_z)](h^{n+1} - h^n) = (\Delta t)\left[\frac{Q^{n+\theta}}{S_s} - (L_{xy} + L_z)h^n\right] \tag{1.80}$$

where I is the identity operator.

Equation 1.80 can be modified by completing the square. One then obtains

$$\{[I - \theta(\Delta t)L_{xy}][I - \theta(\Delta t)L_z)]\}(h^{n+1} - h^n) = (\Delta t)\left[\frac{Q^{n+\theta}}{S_s} - (L_{xy} + L_z)h^n\right] + \theta^2(\Delta t)^2 L_{xy}L_z(h^{n+1} - h^n)$$

where the perturbation term $P \equiv \theta^2(\Delta t)^2 L_{xy}L_z(h^{n+1} - h^n)$ has been added to both sides of equation 1.80. If we assume that the second-order perturbation term is small relative to other terms in the equation, we can write

$$[I - \theta(\Delta t)L_{xy}]Z^n = \Delta t\left[\frac{Q^{n+\theta}}{S_s} - (L_{xy} + L_z)h^n\right] \tag{1.81}$$

$$[I - \theta(\Delta t)L_z](h^{n+1} - h^n) = Z^n \tag{1.82}$$

Note that by neglecting the second-order P term we have committed an additional $O(\Delta t)^2$ error.

Examination of equations 1.81 and 1.82 reveals two equations similar to those we have identified earlier. Equation 1.81 is a two-dimensional flow equation in the areal plane in the unknown Z^n. Equation 1.82 is a one-dimensional flow equation in the vertical dimension in the unknown $(h^{n+1} - h^n)$. The variable Z^n is physically meaningless and cannot be used for any purpose other than to provide input to equation 1.82.

[31] The development followed here is similar to that presented in Celia and Pinder [3].

Let us now expand equation 1.81 and, subsequently equation 1.82 using the definitions of L_{xy} and L_z, that is,

$$\left\{I - \theta(\Delta t)\frac{1}{S_s}\left[\frac{\partial}{\partial x}\left(K_{xx}\frac{\partial}{\partial x}\right) + \frac{\partial}{\partial y}\left(K_{yy}\frac{\partial}{\partial y}\right)\right]\right\} Z^n = \Delta t \frac{Q^{n+\theta}}{S_s}$$
$$- \Delta t \left[\frac{1}{S_s}\left[\frac{\partial}{\partial x}\left(K_{xx}\frac{\partial}{\partial x}\right) + \frac{\partial}{\partial y}\left(K_{yy}\frac{\partial}{\partial y}\right)\right] h^n + \frac{1}{S_s}\frac{\partial}{\partial z}\left(K_{zz}\frac{\partial}{\partial z}\right) h^n\right]$$

We can rearrange this expression to provide a more familiar form. Dividing through by Δt and multiplying through by S_s, we obtain

$$\theta \boldsymbol{\nabla}_{xy} \cdot \mathbf{K} \cdot \boldsymbol{\nabla}_{xy} Z^n = \frac{S_s Z^n}{\Delta t} - W \tag{1.83}$$

where

$$W \equiv Q^{n+\theta} + \boldsymbol{\nabla}_{xy} \cdot \mathbf{K} \cdot \boldsymbol{\nabla}_{xy} h^n + \frac{\partial}{\partial z}\left(K_{zz}\frac{\partial}{\partial z}\right) h^n \tag{1.84}$$

Note that all the information in W is known.

Expanding equation 1.82 and rearranging, we obtain

$$\frac{\theta}{S_s}\frac{\partial}{\partial z}\left[K_{zz}\frac{\partial(h^{n+1} - h^n)}{\partial z}\right] = \frac{h^{n+1} - h^n}{\Delta t} + Y \tag{1.85}$$

$$Y \equiv \frac{Z^n}{\Delta t}$$

where Y is known from the solution of equation 1.83.

The solution of equation 1.83 is a two-dimensional problem similar to that described above for the vertically integrated equations. Equation 1.85, on the other hand, is a one-dimensional problem which was considered in Section 1.10. The importance of creating a series of one- and two-dimensional problems from a single three-dimensional problem is that the former is more computationally efficient. This entire matter will be considered in more depth shortly.

To complete this analysis **we approximate the areal two-dimensional problem using finite elements and the vertical one-dimensional problem using finite differences**. One could also solve the vertical equations using a one-dimensional finite-element approximation.

Consider first the areal equations given by equation 1.83:

$$\theta \boldsymbol{\nabla}_{xy} \cdot \mathbf{K} \cdot \boldsymbol{\nabla}_{xy} Z^n = \frac{S_s Z^n}{\Delta t} - W \tag{1.86}$$

We first approximate Z^n using a finite series of the form

$$Z^n \approx \hat{Z}^n = \sum_{j=1}^{J} Z_{nj}\phi(x, y)_j \tag{1.87}$$

where $\phi(x, y)_j$ is a two-dimensional basis function. In *PTC*, basis functions defined over triangular, rectangular, and isoparametric elements are employed. We address the matter of alternative basis functions later. For now, let us proceed formally with the definition of a residual formulation for equation 1.86.

Applying Galerkin's method to equation 1.86, we obtain

$$\int_{\Omega} \left(\theta \nabla_{xy} \cdot \mathbf{K} \cdot \nabla_{xy} \hat{Z}^n - \frac{S_s \hat{Z}^n}{\Delta t} + W \right) \phi(x, y)_i \, d\Omega = 0, \qquad i = 1, \ldots, I \tag{1.88}$$

which upon modification using Green's theorem yields

$$\int_{\Omega} \left[-\theta \mathbf{K} \cdot \nabla_{xy} \hat{Z}^n \cdot \nabla_{xy} \phi(x, y)_i - \left(\frac{S_s \hat{Z}^n}{\Delta t} - W \right) \phi(x, y)_i \right] d\Omega$$
$$+ \int_{\partial\Omega} \mathbf{K} \cdot \frac{\partial Z^n}{\partial \mathbf{n}} \phi(x, y)_i \, dl = 0, \quad i = 1, \ldots, I \tag{1.89}$$

The question naturally arises as to how to apply boundary conditions to equation 1.89. To see how this is done, we rewrite the definition of Z^n, that is,

$$Z^n = [1 - \theta(\Delta t) L_z](h^{n+1} - h^n) \tag{1.90}$$

Note that if we have a constant head boundary (Dirichlet or type 1), $h^{n+1} = h^n$; that is, the head does not normally change with time. If it does change with time, the relationship is known. In either case Z^n is known from equation 1.90, and therefore equation 1.89 can be solved using the standard methods that have been and will be discussed.

In the case of a Dirichlet-condition specification, the last term on the left-hand side of equation 1.89 is not important, since the equation is eliminated at the Dirichlet node. However, in the case of a flow (Neumann or second-type) boundary condition, the equation in which the flow is specified remains and must be solved. The last term on the left-hand side of equation 1.89 must be specified, since it represents the flow boundary condition. Notice that this term contains the normal derivative of Z^n, that is,

$$\frac{\partial Z^n}{\partial \mathbf{n}} = \frac{\partial}{\partial \mathbf{n}} [(I - \theta(\Delta t) L_z)(h^{n+1} - h^n)] \tag{1.91}$$

If the derivative

$$K_{zz} \frac{\partial}{\partial z} (h^{n+1} - h^n)$$

is specified, the second derivative, that is,

$$\frac{\partial}{\partial z} \left[K_{zz} \frac{\partial}{\partial z} (h^{n+1} - h^n) \right]$$

is zero. Thus this term vanishes.

In summary, when a **Dirichlet condition** is specified, it is easily accommodated. When a **Neumann condition** is specified on flow, the flux term involving Z^n vanishes. The actual flow boundary condition will arise naturally out of the evaluation of W.

Let us now consider equation 1.85, that is,

$$\theta \frac{\partial}{\partial z}\left[K_{zz}\frac{\partial(h^{n+1}-h^n)}{\partial z}\right] = S_s\frac{h^{n+1}-h^n}{\Delta t} + S_s\frac{Z^n}{\Delta t} \tag{1.92}$$

Application of a standard finite-difference formulation yields

$$\frac{K|_B[(h_{i+1}^{n+1}-h_i^{n+1})/\Delta z_B] - K|_A[(h_i^{n+1}-h_{i-1}^{n+1})/\Delta z_A]}{\Delta z_i} = \frac{S_s}{\theta}\frac{h_i^{n+1}-h_i^n}{\Delta t} + \frac{S_s}{\theta}\frac{Z^n}{\Delta t} \tag{1.93}$$

where, as defined earlier,

$$\frac{K_B}{\Delta x|_B} = \frac{[K_{i+1}/(\Delta x_{i+1}/2)][K_i/(\Delta x_i/2)]}{[K_{i+1}/(\Delta x_{i+1})/2] + [K_i/(\Delta x_i)/2]} \tag{1.94}$$

and the ratio $K_A/\Delta x|_A$ is defined in an analogous fashion.

The procedure for the solution of equations 1.89 and 1.93 is the following. The two-dimensional problem defined by equation 1.89 is solved for each layer of the model, one layer at a time, using the last calculated values for information on adjacent layers. When this is completed for all layers, equation 1.93 is solved for all of the vertical lines of nodes, one line at a time. There will be one such line of nodes for each node appearing in the two-dimensional finite-element net. This two-step process will be repeated for each time step.

Although the procedure above appears attractive, especially from the point of view of error analysis, it is not exactly the same as what is coded in *PTC*. In the following section, we arrive at a similar result from a quite different point of view.

As our point of departure, consider the three-dimensional groundwater-flow equation

$$\nabla \cdot \mathbf{K} \cdot \nabla h = S_s\frac{\partial h}{\partial t} + Q \tag{1.95}$$

which we can write in weighted residual form as

$$\int_\Omega \left[(-\mathbf{K}\cdot\nabla\hat{h})\cdot\nabla\phi(\mathbf{x})_i - \left(S\frac{\partial\hat{h}}{\partial t} + Q\right)\phi(\mathbf{x})_i\right] d\Omega$$

$$+ \int_{\partial\Omega} \mathbf{K}\cdot\frac{\partial h}{\partial \mathbf{n}}\phi(\mathbf{x})_i \, dl = 0, \qquad i = 1, \ldots, I \tag{1.96}$$

Let us now define the approximating function $\hat{h}$ as

$$\hat{h}(x, y, z, t) = \sum_{j=1}^{J} h(z, t)_j \phi(x, y)_j \tag{1.97}$$

where we now have defined our unknown coefficients $h(z, t)_j$ as functions of both time and the vertical dimension z. The basis functions are now a function of the two coordinates in the areal, or, generally speaking, horizontal plane.

We now substitute equation 1.97 into equation 1.96, to obtain

$$\begin{aligned}
&\int_{\Omega} \left\{ -\mathbf{K} \cdot \boldsymbol{\nabla} \left[\sum_{j=1}^{J} h(z, t)_j \phi(x, y)_j \right] \right\} \cdot \boldsymbol{\nabla} \phi(\mathbf{x})_i \, d\Omega \\
&- \int_{\Omega} \left(S_s \frac{\partial \left[\sum_{j=1}^{J} h(z, t)_j \phi(x, y)_j \right]}{\partial t} + Q \right) \phi(\mathbf{x})_i \, d\Omega \\
&+ \int_{\partial\Omega} \mathbf{K} \cdot \frac{\partial h}{\partial \mathbf{n}} \phi(\mathbf{x})_i \, dl = 0, \qquad i = 1, \ldots, I
\end{aligned} \tag{1.98}$$

Next we expand the differentials, taking into account the fact that the basis functions now are a function only of (x, y), and the coefficients are a function of (z, t).

$$\begin{aligned}
&\int_{\Omega_{xy}} \left\{ -\mathbf{K} \cdot \boldsymbol{\nabla}_{xy} \left[\sum_{j=1}^{J} h(z, t)_j \phi(x, y)_j \right] \right\} \cdot \boldsymbol{\nabla}_{xy} \phi(x, y)_i \, d\Omega \\
&+ \int_{\Omega_{xy}} \frac{\partial}{\partial z} \left[K_{zz} \frac{\partial}{\partial z} \sum_{j=1}^{J} h(z, t)_j \right] \phi(x, y)_j \phi(x, y)_i \, d\Omega \\
&- \int_{\Omega_{xy}} \left(S_s \frac{\partial \left[\sum_{j=1}^{J} h(z, t)_j \phi(x, y)_j \right]}{\partial t} + Q \right) \phi(x, y)_i \, d\Omega \\
&+ \int_{\partial\Omega_{xy}} \mathbf{K} \cdot \frac{\partial h}{\partial \mathbf{n}} \phi(x, y)_i \, dl = 0, \qquad i = 1, \ldots, I
\end{aligned} \tag{1.99}$$

where the derivative in the z-coordinate direction has been modified through integration by parts. This equation can be rearranged, after again employing integration by parts to the derivatives in z, to give

$$\begin{aligned}
&\sum_{j=1}^{J} h(z, t)_j \int_{\Omega_{xy}} [-\mathbf{K} \cdot \boldsymbol{\nabla}_{xy} \phi(x, y)_j] \cdot \boldsymbol{\nabla}_{xy} \phi(x, y)_i \, d\Omega \\
&+ \frac{\partial}{\partial z} \left[K_{zz} \frac{\partial}{\partial z} \sum_{j=1}^{J} h(z, t)_j \right] \int_{\Omega_{xy}} \phi(x, y)_j \phi(x, y)_i \, d\Omega
\end{aligned}$$

$$
- \sum_{j=1}^{J} \frac{\partial \left[h(z,t)_j \right]}{\partial t} \int_{\Omega_{xy}} [S_s \phi(x,y)_j] \phi(x,y)_i \, d\Omega
$$

$$
+ \int_{\Omega_{xy}} Q \phi(x,y)_i \, d\Omega
$$

$$
+ \int_{\partial\Omega_{xy}} \mathbf{K} \cdot \frac{\partial h}{\partial \mathbf{n}} \phi(x,y)_i \, dl = 0, \qquad i = 1, \ldots, I \tag{1.100}
$$

To accommodate the z and time derivatives, a finite-difference approximation is introduced, that is,

$$
\left. \frac{\partial h(z,t)_j}{\partial t} \right|_k \approx \frac{h_j^{n+1k} - h_j^{nk}}{\Delta t} \tag{1.101}
$$

$$
\left. \frac{\partial}{\partial z} \left[K_{zz} \frac{\partial}{\partial z} h(z,t)_j \right] \right|_{n+1}
$$

$$
\approx \frac{K|_B[(h_j^{n+1k+1} - h_j^{n+1k})/\Delta z_B] - K|_A[(h_j^{n+1k} - h_j^{n+1k-1})/\Delta z_A]}{\Delta z_i} \tag{1.102}
$$

where the parameters defined at A and B are defined as in equation 1.94. Substitution of equations 1.101 and 1.102 into 1.100 yields

$$
\sum_{j=1}^{J} h_j^{n+1k} \int_{\Omega_{xy}} [-\mathbf{K} \cdot \boldsymbol{\nabla}_{xy} \phi(x,y)_j] \cdot \boldsymbol{\nabla}_{xy} \phi(x,y)_i \, d\Omega|_k
$$

$$
+ \sum_{j=1}^{J} \frac{K|_B[(h_j^{n+1k+1} - h_j^{n+1k})/\Delta z_B] - K|_A[(h_j^{n+1k} - h_j^{n+1k-1})/\Delta z_A]}{\Delta z_i}
$$

$$
\int_{\Omega_{xy}} \phi(x,y)_j \phi(x,y)_i \, d\Omega|_k - \sum_{j=1}^{J} \frac{h_j^{n+1k} - h_j^{nk}}{\Delta t}
$$

$$
\int_{\Omega_{xy}} [S_s \phi(x,y)_j + Q] \phi(x,y)_i \, d\Omega|_k + \int_{\partial\Omega_{xy}} \mathbf{K} \cdot \frac{\partial h}{\partial \mathbf{n}} \phi(x,y)_i \, dl|_k = 0,
$$

$$
i = 1, \ldots, I \tag{1.103}
$$

Equation 1.103 can now be written in matrix form for the kth layer as

$$
[A]_k \frac{\{h^{n+1k} - h^{nk}\}}{\Delta t} = [B]_k \{h^{n+1k}\} + \{F\}_k + \{g\}_k \tag{1.104}
$$

where

$$a_{ij}^k = \left[\int_{\Omega_{xy}} S_s \phi(x, y)_j \phi(x, y)_i \, d\Omega \right]_k \tag{1.105}$$

$$b_{ij}^k = \left[\int_{\Omega_{xy}} [-\mathbf{K} \cdot \boldsymbol{\nabla}_{xy} \phi(x, y)_j] \cdot \boldsymbol{\nabla}_{xy} \phi(x, y)_i d\Omega \right]_k \tag{1.106}$$

$$f_i = [C]^{k+1} \{h^{n+1k+1}\} + [C]^k \{h^{n+1k}\} + [C]^{k-1} \{h^{n+1k-1}\} \tag{1.107}$$

and where, for example,

$$c_{ij}^k = \frac{(K|_B) / \Delta z_B + (K|_A) / \Delta z_A}{\Delta z_i} \int_{\Omega_{xy}} \phi(x, y)_j \phi(x, y)_i d\Omega \tag{1.108}$$

$$g_i = \left[\int_{\partial\Omega_{xy}} \mathbf{K} \cdot \frac{\partial h}{\partial \mathbf{n}} \phi(x, y)_i \, dl|_k - \int_{\Omega_{xy}} Q\phi(x, y)_i \, d\Omega \right]_k \tag{1.109}$$

It has been found through experience that there are advantages to **lumping** some matrices. Lumping is the process by which matrices are **diagonalized** by summing their row values. Specifically, the matrices associated with the time derivative and the z-direction derivative are cases in point. If this procedure is applied to the time-derivative matrix, which is often called the **mass matrix** due to the fact that the computational roots of the finite-element method are in engineering mechanics, one obtains

$$a_{ii}^k = \sum_{j=1}^{J} a_{ij}^k \tag{1.110}$$

And for the z-direction derivative

$$c_{ii}^\alpha = \sum_{j=1}^{J} c_{ij}^\alpha \tag{1.111}$$

All off-diagonal elements in $[A]$ and $[C]$ are defined as zero.

Let us define the new diagonal matrices as $[A_D]$ and $[C_D]$. Given the new definitions of these matrices, we obtain

$$[A_D]^k \frac{\{h^{(n+1)k} - h^{nk}\}}{\Delta t} = [B]^k \{h^{n+1,k}\} + [C_D]^{k+1} \{h^{(n+1)(k+1)}\} + [C_D]^k \{h^{(n+1)k}\} + [C_D]^{k-1} \{h^{(n+1)(k-1)}\} + \{g\}^k \tag{1.112}$$

The two-step *PTC* algorithm can now be written.

Step 1

$$[A_D]^k \frac{\{h^{(n+1)k} - h^{nk}\}}{\Delta t} = [B]^k\{h^{(n+1)k}\} + [C_D]^{k+1}\{h^{nk+1}\} + [C_D]^k\{h^{nk}\} + [C_D]^{k-1}\{h^{nk-1}\} + \{g\}^k \tag{1.113}$$

By lagging the vertical derivative [i.e., holding it at the old time level n while calculating an intermediate solution for the $n + 1$ level, namely $(n + 1)^*$], we have uncoupled the horizontal equations in each layer one from the other. Thus, the solution of equation 1.112 for all of the layers, $k = 1, \ldots, M$, consists of solving the horizontal equations M times, once for each layer. Equation 1.113 is comparable to, but somewhat different from, equations 1.83 and 1.84 which we developed formally earlier by completing the square.

Step 2 Having obtained the intermediate solution, we now perform vertical sweeps. The appropriate equation for the vertical sweep is

$$[A_D]^k \frac{\{h^{(n+1)k} - h^{nk}\}}{\Delta t} = [B]^k\{h^{(n+1)k}\} + [C_D]^{k+1}\{h^{(n+1)(k+1)}\} + [C_D]^k\{h^{(n+1)k}\} + [C_D]^{k-1}\{h^{(n+1)(k-1)}\} + \{g\}^k, \qquad k = 1, \ldots, M \tag{1.114}$$

Note that if we had not lumped the C matrix, equation 1.114 would not be a one-dimensional problem because the nodes connected horizontally to the ith node would still contain unknowns such that the system of equations would remain three-dimensional. Equation 1.114 is once again comparable to but different from equation 1.85 developed above by completing the square.

A simplification of the system of equations above is achieved by subtracting 1.114 from 1.113, which yields

$$[C_D]^{k+1}\{h^{(n+1)(k+1)}\} + [C_D]^k\{h^{(n+1)k}\}[C_D]^{k-1}\{h^{(n+1)(k-1)}\} - [A_D]^k \frac{\{h^{(n+1)k}\}}{\Delta t} = [C_D]^k\{h^{n(k+1)}\} + [C_D]^k\{h^{nk}\}[C_D]^{k-1}\{h^{n(k-1)}\} - [A_D]^k \frac{h^{(n+1)k}}{\Delta t}, \qquad k = 1, \ldots, M \tag{1.115}$$

One can now solve equation 1.115 for each vertical array of nodes in the three-dimensional model. **The overall protocol then becomes one of solving for each layer of nodes, one at a time, followed by a solution of each vertical line of nodes, one at a time**. The completion of the two steps constitutes the calculations necessary for one time step. **The resulting solution is fully three-dimensional**.

As noted earlier, the two-dimensional problems can be solved independently of one another. Therefore, several layers can be solved simultaneously on a parallel-processing computer. Similarly, each of the vertical sweeps can be solved independently so these equations can also be solved efficiently on a parallel-processing computer.

The system of equations 1.113 that arises out of the set of two-dimensional problems is solved with a standard Gaussian elimination equation solver designed specifically for symmetric matrices. The one-dimensional problem generates a tridiagonal matrix that is efficiently solved using the Thomas algorithm, which is a special case of Gaussian elimination designed for tridiagonal matrices.

1.16.2 Bandwidth

The amount of work required to solve a matrix equation using standard Gaussian elimination is proportional to nb^2, where n is the number of equations and b is the matrix bandwidth. Thus the matter of the bandwidth is important. To examine this issue, consider the finite-element net shown in Figure 1.58. The bandwidth is obtained by first subtracting the largest node number from the smallest node number in each element. The largest of these computed values, after considering every element, is now recorded. The **bandwidth** is obtained by multiplying this number by 2 and adding 1.

The top example in Figure 1.58 is numbered in the long dimension and has a bandwidth of 35. The lower net is numbered across the short dimension and has a bandwidth of 9. The lesson is obvious: One numbers nodes in the short dimension to minimize bandwidth. The bandwidth for a three-dimensional mesh is obtained by taking the difference between the eight node numbers that define each cubic element.

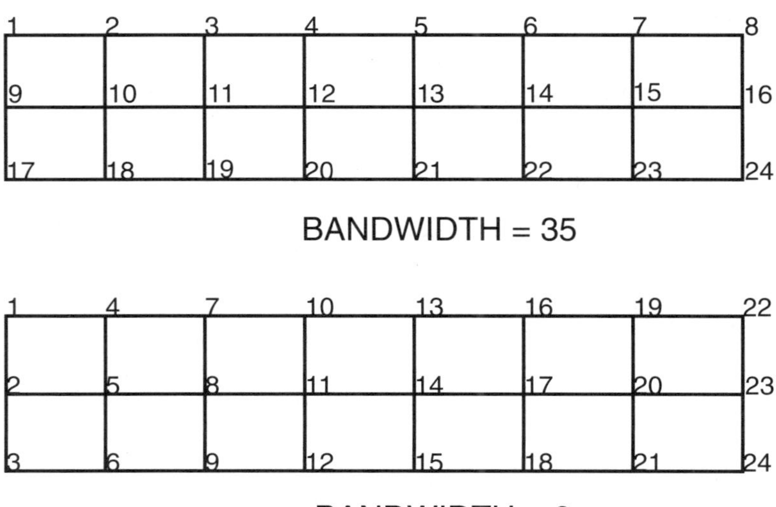

FIGURE 1.58. *Nodal-numbering arrangements that illustrate the importance of numbering in the short dimension to minimize bandwidth.*

The result is enormous. Thus one sees the advantage of solving each layer separately, such as *PTC* does, rather than all the layers at once.

Argus ONE has the ability to minimize the bandwidth of the resulting coefficient matrices by optimally numbering the nodes. To do this make the *Mesh* layer active, select *Special*, then *Renumber*, then *Optimize Bandwidth*.

1.16.3 Running PTC

We now are at the point of submitting our data set to *PTC* and initiating the simulation. However, before we can do that we must return to the *PTC* configuration window and enter some of the information we deferred including in earlier sections. From Figure 1.59 we see that we have yet to address the *Steady state criterion* text box. The value placed in this box is the maximum change between time-step calculations that will qualify the solution as having reached steady-state flow. We must also push the appropriate *Mesh type* button. If left unchecked, a triangular-element configuration will be assumed.

Now select the *Stresses* tab. The resulting window is shown in Figure 1.60. To activate the flow-simulation option in *PTC*, one must complete certain of the items appearing in the *General control* boxes. The first step is to turn on the check box for *Do flow*. The *Use memory* box should also be turned on. If a velocity calculation based on the head simulation is desired, turn on the *Do velocity* check box.

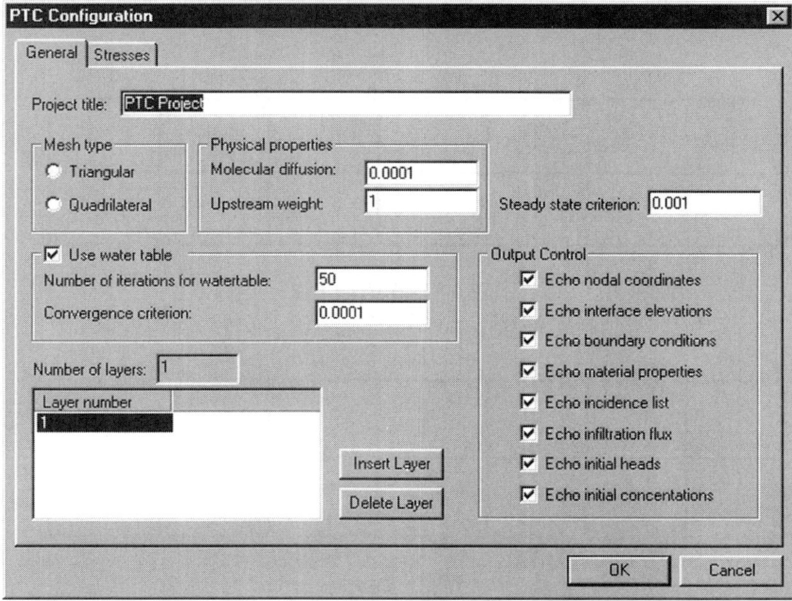

FIGURE 1.59. *PTC Configuration dialog box opened by selection of New PTC project or via the PIEs option from the menu bar.*

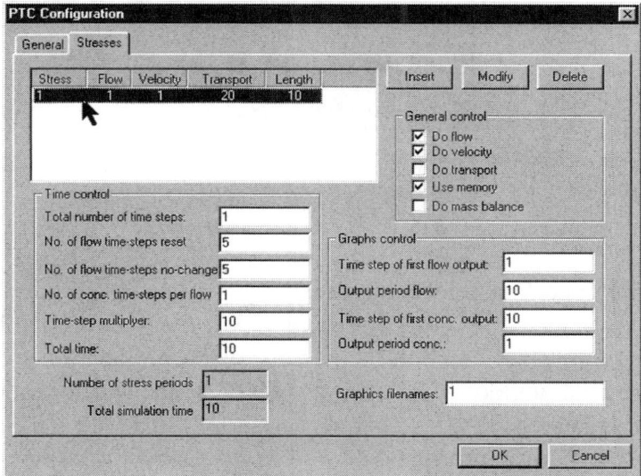

FIGURE 1.60. *Multiple-stress period information is provided via this window along with various other project information.*

Consider next the *Time control* series of text boxes. In the *Total time* check box, provide the total length of time for the simulation in the time units selected earlier to define the aquifer parameters. Next, place in the *Total number of time steps* text box the number of time steps you wish to use to simulate the total time. The division of the total time by the number of time steps will define the time step Δt, provided that it remains constant throughout the simulation. However, especially in flow calculations, it may be advantageous to increase the size of the time step as the groundwater system approaches steady state. This is achieved by placing a time-step multiplier in the *Time-step multiplyer* text box. If the analyst wishes to modify the time step after every n time steps, then n is placed in the *No. of flow time-steps reset* text box. On the other hand, if the goal is to stop the time-step growth after m time steps, then m should be placed in the *No. of flow time-steps no-change* text box.

Turning now to the *Graphs control* text boxes, we first indicate the first time step for which we wish to save to disk our flow-simulation solution for use in presenting graphical solutions. The appropriate time step is placed in the *Time step of first flow output* text box. In the *Output period flow* text box is placed the number of time steps that are to elapse between successive solution outputs to disk.

Note that the information above is to be provided for each stress period. In other words, different time control and general control profiles can be provided for each stress period.

Tucson Example

Now that the *PTC Configuration* window is complete, all the auxiliary conditions have been provided along with the stresses and aquifer parameters we are prepared to launch the simulator. The procedure to do so involves calling *PTC* from Argus

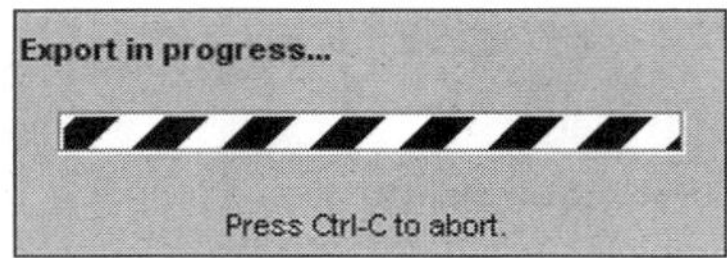

FIGURE 1.61. *The transfer of data from Argus ONE to the DOS window running PTC is indicated by this barber-pole window.*

ONE. To do this, one first selects the *Mesh* layer. Then, from the *PIEs* option on the menu bar one selects *Run PTC*. An *Enter export file name* dialog box appears in which you can place a file name to be used for recording and transferring the *PTC* input files to the DOS window. The default is **PTC _Mesh.exp**. Upon selecting *Save*, the *PTC* model is launched.

During the transfer of information from Argus ONE to the DOS window, you will see a barber-pole icon such as shown in Figure 1.61. Upon completion of the transfer, *PTC* begins executing and a DOS screen appears that shows the progress of the simulation. In Figure 1.62 the successful completion of the first time step for flow is indicated. In addition, we are informed that after the first time step the criteria set for identifying a steady-state flow solution has been realized. The next four lines of DOS screen output indicate that two flow and transport time steps have been completed successfully, as has the third time step for flow. As the simulation progresses, the completed time steps are indicated until the entire simulation period has been accommodated.

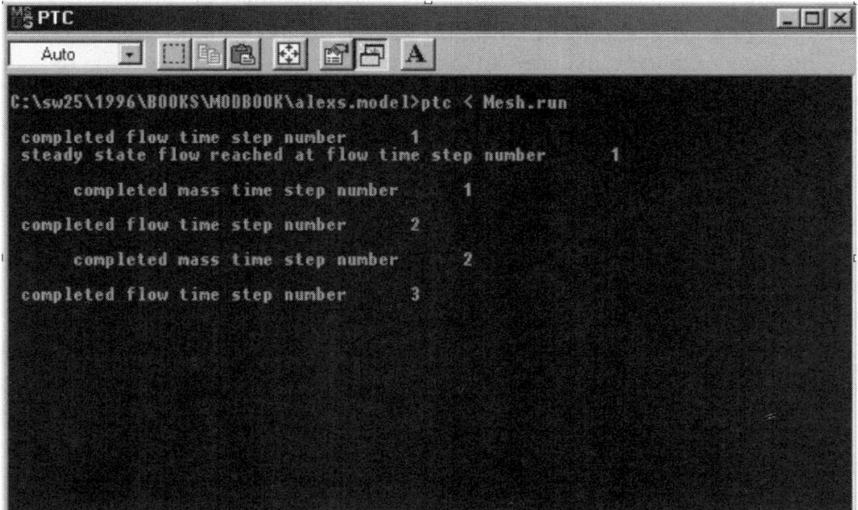

FIGURE 1.62. *DOS screen indicates the progress of the PTC simulation. If there are errors in the input or the array space allocated for the simulation is exceeded, the concomitant error will appear on this screen.*

1.17 OUTPUT

There are a number of ways to view the **hydraulic-head output**. The presentation of an array of numbers corresponding to nodal locations is the most primitive approach and is of limited utility. However, this kind of information is provided in the **ptc.out** file and may be useful for debugging input data and for transferring data to graphics programs other than Argus ONE.

In viewing the graphics output files, the following protocol is used. The name provided in the *PTC Configuration*, as indicated in Figure 1.60, is used to record the hydraulic head values generated by *PTC* according to the *Graphs Control* specifications. Let us assume that you select *flow*. Attached to this name there will be *_s1.1*, yielding *flow_s1.1*. Because of the limitation of the DOS file name to eight characters, the use of the name *flowonly* would not be acceptable since the total name would then be 11 characters. The *s1* refers to the fact that you are using only one stress period, and this is the output for it. The last *1* in the name indicates that this is the first output. Keep in mind that the first output may not coincide with the first time step, depending on the *Graphs control* specifications.

For example, let us assume that you set your file name to **flow**, set the times steps to 20, and set the graphics control output for flow to 5, starting at time 5. Four files would be created at time steps 5, 10, 15, and 20. The names would be *flow_s1.1, flow_s1.2, flow_s1.3*, and *flow_s1.4*.

Contours of hydraulic head are of great utility inasmuch as they permit visualization of the pattern of head values. Such patterns provide insight into the direction and magnitude of groundwater flow because flow directions are normally orthogonal to the head contours and the magnitude of the flow is inversely proportional to the distance between successive contours, given the contour interval is constant. Three-dimensional surfaces of the head (fishnet plots) provide a striking picture of the fluid potential surface and are useful for visual presentation. However, they are not very useful for quantitative analysis.

A variant on the contouring and the three-dimensional surfaces is the addition of color to indicate the magnitude of the head. Color contours or three-dimensional surfaces that change color with elevation are available. A sequence of head surfaces computed for different time values can be concatenated together to produce the equivalent of an animated movie. The dynamics of the system are thereby visualized. In the case of flow, however, the dynamics are generally very rapid and there is relatively little movement to observe. This is not true of contaminant transport, however. In this case the movement of the contaminants is quite slow and animation can be very effective.

Tucson Example

If the DOS window that reports on the progress of the simulation indicates successful completion of the simulated period of analysis, the time has come to view the model output. As noted Argus ONE can provide a graphical representation of the solution that has been generated. The process of retrieving and eventually viewing the output

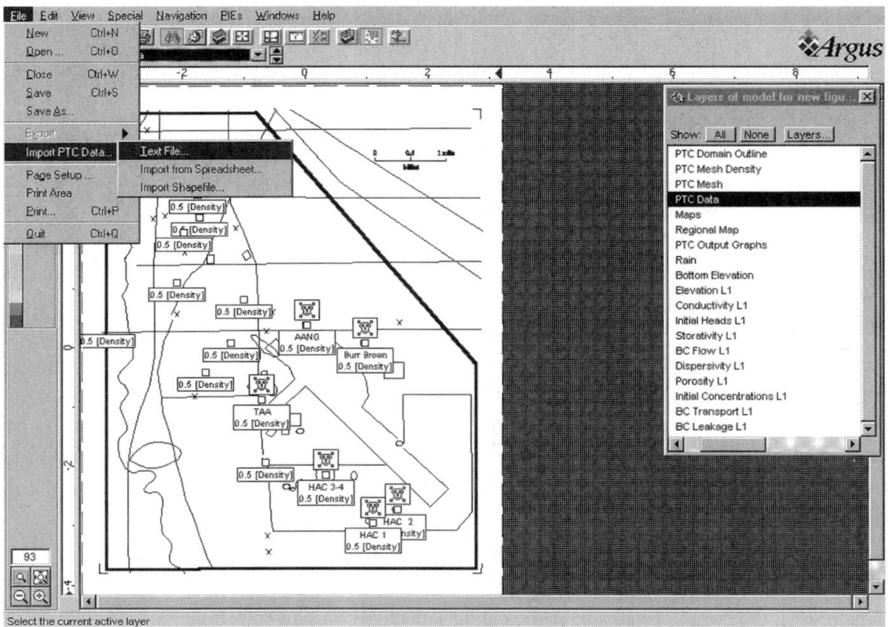

FIGURE 1.63. *Command sequence required to import data for graphical presentation of the PTC data file that contains the results of the flow simulation.*

requires several steps. The first step retrieves the output file you wish to examine. In other words, you must select the time step for which you wish to view the solution. To do this, we first activate the *PTC Data* layer. From the *File* option on the menu bar, select *Import PTC Data* and then *Text File* (Figure 1.63).

Upon completion of the sequence of steps above, an *Input Data* window opens. The default values in this window must be changed to import the *PTC* output file. The

Import Data
Please specify the type of data to import:
OK
Cancel
Scattered data
Mesh data:
Read triangulation from file...
Elements type: Triangular
Read triangulation from layer: Mesh
Grid data:
Read triangulation from file:
Grid is: Block Centered Line Centered
Read triangulation from layer: <None>

FIGURE 1.64. *Input Data format window used to inform Argus ONE of the format of the file that contains the computed results of the simulation.*

Mesh data button and the *Read triangulation from layer* button must be activated. The resulting window is shown in Figure 1.64.

Acceptance of the information in the *Input Data format* window activates the *Choose file to import* window. Listed here are the various files generated for *PTC* by Argus ONE and the resulting output files. To access them, the *Files of type* option must read *All Files (*.*)*. Files of the form described above will appear. For example, *flow_s1.1* contains the head values created during the fifth time step of the example simulation described above. Select the file you wish to view graphically and click on *Open*.

Upon completion of the sequence of steps above the window presented in Figure 1.65 appears. Select a *Maps* layer. From the toolbar that appears in the upper left-hand corner of this window, click and hold on the *Post-Processing tool* (the darkened tool shown in Figure 1.65). A list of available options appears. To generate a contoured output, the fourth tool from the top (again darkened in Figure 1.65) must be selected. The remaining tools generate other forms of visual output which we do not consider here but are described in the Argus ONE documentation.

Once the *Create a contour map* tool is selected, a crosshair-shaped cursor is created. Place the cursor anywhere on the active-window area and draw a rectangle. The size of the rectangle is irrelevant for our purposes.

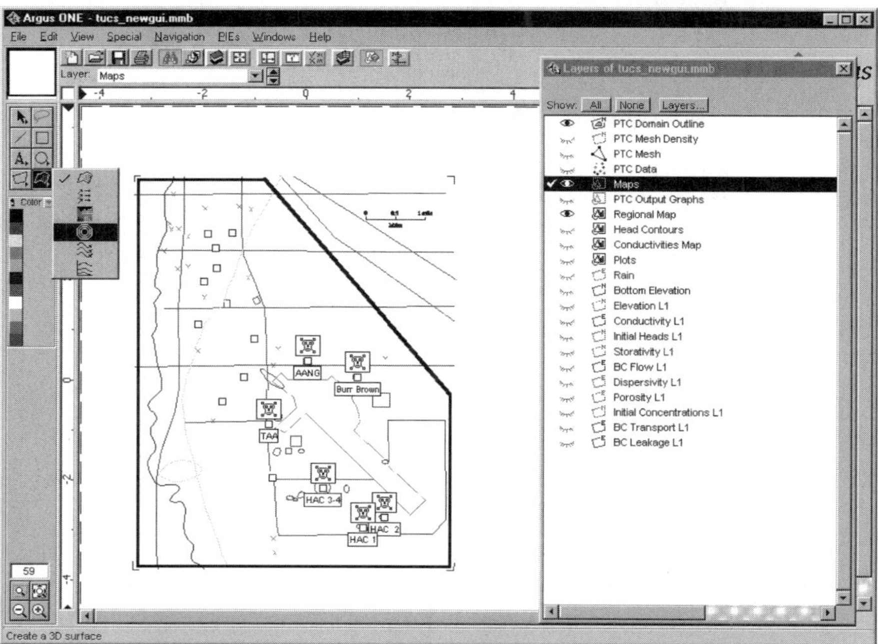

FIGURE 1.65. *Procedure used to select the graphical format to be used in presentation of the selected PTC output file.*

Completion of the box activates the *Contour Diagram* window illustrated in Figure 1.66. The default value for the *Layer* option is not the one to be used for contouring the head values output from the model. Rather, the layer *PTC Data* must be selected, as shown in Figure 1.66. Default values for the remaining parameters should be selected.

Now select the *Position tab* on the *Contour Diagram* window. Select the *Overlay Source Data* option as illustrated in Figure 1.67. By selecting *OK* in this window, the contour is generated as shown in Figure 1.68. The values of the contours are indicated by the scale that appears to the right of the modeled area. On the computer screen the contours are in color and their values easily identified. By selecting the *Titles* tab on the *Contour Diagram* window, a suitable title can be displayed at the top of the contoured output.

It may be convenient, when preparing figures for presentation, to have the ability to create layers specifically for that purpose. You may, for example, wish to have a layer that contains a site map or a contour map of the elevation of the top of a geological unit. A new layer is created by selecting *View*, the *Layers*, then *New* from the *Layer* frame. The *Layers* window can also be activated using the *Layers Dialog*

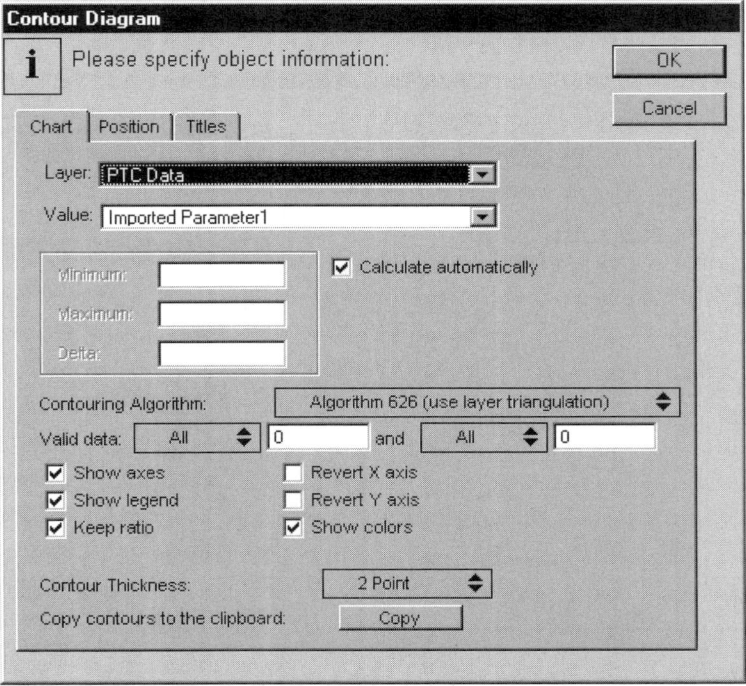

FIGURE 1.66. *Contour Diagram window used to input data to be contoured. Note that the default for this window must be changed to appear as illustrated above. More specifically the layer and value list box values must be changed.*

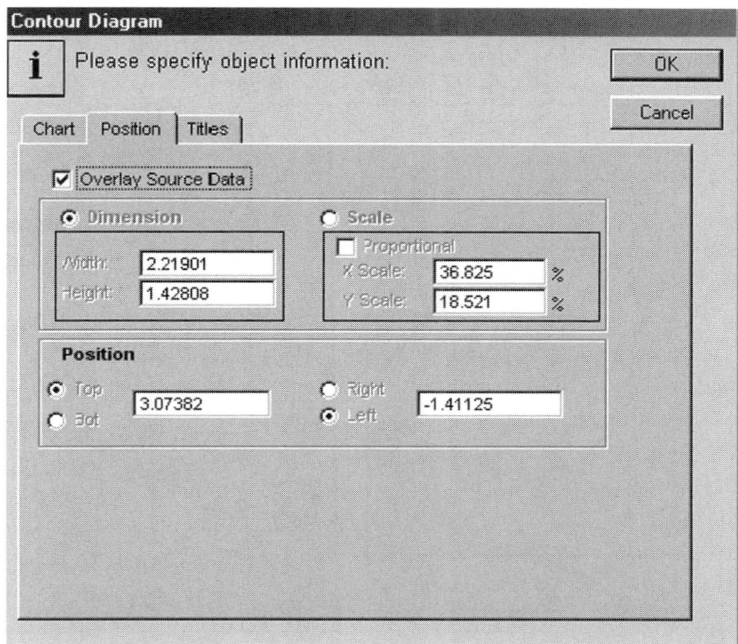

FIGURE 1.67. *Selection of the Position tab on the Contour Diagram window reveals the above window. The default value of the Overlay Source Data check box must be changed to activate this option.*

button. A new layer called *New Layer 1* with type *Information* will appear. Click the spin button to the right of the entry *Information* and from the resulting drop down list select *Maps*. Change the entry *New Layer 1* to reflect the type of graphical material that will be stored there. For example, you might select *Top of Clay Elevation Map* to reflect the fact that you will be placing contour maps of the top of clay elevation in that layer for easy reference.

To create a plot of the elevation of the top of the clay layer, the *Top of Clay Elevation Map* layer is made active. The *Post-Processing* tool is selected as was done in the presentation of the head output from *PTC* and the cross-shaped cursor is used to draw a box somewhere in the work area. The *Contour Diagram* dialog box will appear and the default entries for *Layer*: will be *PTC Mesh* and for *Value*: will be *Bottom Elevation*. Click the spin button associated with *Value* and from the drop-down list select the layer associated with the top of clay elevation which we will call *Clay Tops*. From this point forward, the graphing protocol is as described above for head values.

Argus ONE permits the manipulation of information to facilitate the visualization of quantities other than state-variable output. If, for example, one wished to contour the depth of water above the clay layer, it would be necessary to subtract the head or water-table surface in the top layer from the underlying clay surface and create a new surface composed of this difference.

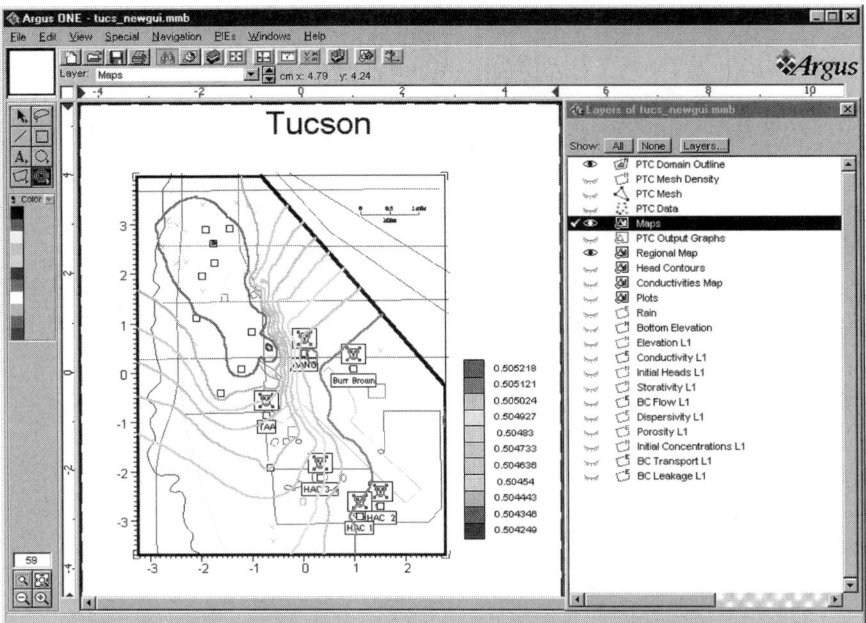

FIGURE 1.68. *Contoured head values for the Tucson model.*

The first step in this process is to create a new layer, let us call it *Water Depth*. The procedure for making the layer was described above. We select as our layer type, the default value *Information*. A *Layer Parameters* dialog box is revealed containing the name *Water Depth*. Click on the f_x button and write in the resulting *Expression* box the relationship *PTC Data-Clay Tops*.

To complete the process, it is necessary to make the information on the clay layer elevation available at the nodes where the water levels are provided. We first activate the *PTC Mesh* layer from the *Layers* window which is reached following the protocol described above. By double-clicking to the left of the open eye a check mark should appear. In the lower part of the window a dialog box labeled *Layer Parameters* is revealed. Move to the bottom of the table and click on the last entry. Now click on the *New* button to the right of the parameter list. A new parameter will be created appropriately called *New Parameter*. Let us rename it as *Clay Tops Too*. Click the f_x that appears to the right of the new parameter. The *Expression* editor will appear. Enter *Water Depth*.

Now create the maps layer that will be used to plot the *Water Depth* contours. Let's call it *Water Depth Plot*. Open the *Water Depth Plot* layer, proceed to the *Contour Diagram* dialog box, select *Layer: PTC Mesh*, and finally select *Clay Tops Too*. At this point contours will appear that describe the water thickness.

Another useful concept is that of changing the units of the state variable prior to presentation. The use of the logarithm of concentration rather than concentration itself as the variable to be contoured is an example. In this case the procedure described above is modified such that when the *Expression* editor opens in response to the creation of a *New Parameter* in a new layer, and the f_x button is pushed, the *Mathematical* option is selected from the list box of *Functions:*. In the case of our example, *Log* or *Log10* is selected from the list of options by double clicking the choice of options depending upon whether one wishes to plot the natural or base 10 logarithm of concentration respectively. Where the word *number* appears as the argument of the logarithm chosen one places the name *PTC Data* layer which is assumed to contain the selected state variable (concentration). The new layer will now contain the logarithm of the state variable, that is the logarithm of concentration. The new layer can ow be plotted via a new maps layer as was the case in the *Water Depth* example above.

1.18 CALIBRATION

Calibration involves modification of the parameters, boundary conditions, and forcing functions in the model in such a way as to assure that the model reproduces observed-head values. Although this aspect of modeling is very important, relatively little has been written about it. A recent exposition on the subject is provided by Hill [16] in a brief paper and in a more comprehensive report [17]. Much of the following material is drawn from those documents.

1.18.1 Model Building Guidelines

1. *Apply the principle of parsimony.* The model should evolve from the simple to the complex. Heterogeneity, for example, should be added only as evidence surfaces that a homogeneous model is either unable to reproduce observed hydraulic-head values or geological information is found to be in conflict with the homogeneity hypothesis. In general, for flow modeling, local changes in hydraulic conductivity have a disproportionately small impact on the head solution. Inasmuch as the flow equation is of the heat equation type, it tends to smooth the head solution, thereby reducing the influence of relatively small zones of differing hydraulic conductivity.

   The impact of small zones of variable hydraulic conductivity is much greater in the case of mass transport, where the velocity is affected markedly by hydraulic-conductivity changes. In contrast to the flow equation, the transport equation behaves as a hyperbolic equation when convective transport dominates dispersive or diffusive transport. In this case there is little smoothing of the solution. Especially important are areas of high hydraulic conductivity in a formation characterized otherwise by low hydraulic-conductivity material.
2. *Use a broad range of information to constrain the problem.* In general, there are a number of measurable quantities that can be used for calibration of a model: net infiltration from rainfall, contributions or reductions to streamflow

discharge due to groundwater infiltration or exfiltration, anthropogenic factors such as losses from pipes, or loss and gain from sewers, for example. Occasionally, information provided by witnesses can be helpful. Observations regarding the behavior of surface-water reservoirs as documented in reports, depositions, or court testimony may provide input into the probable rate of infiltration to the underlying aquifer.

In addition, there is information available from well logs, boring logs, or cone penetrometer measurements that can provide conceptual input and constraints on the lithology of aquifers and confining layers. It is often useful to create cross sections and fence diagrams, if they do not already exist, to facilitate determination of the most probable topology (tops and bottoms) for various geological units of hydrogeological importance. Insights provided by structural geologists may provide valuable insight into large-scale geological features such as anticlines, synclines, dikes, and faults. Geophysical information can be helpful in determining the depth to water, location of sources such as metal tanks, and the interfaces between geological units. In the end the goal is to use as much information as is available to produce the most self-consistent, conceptually simple, and physically reasonable representation of the geohydrological system possible.

3. *Include as many kinds of measurable data as practicably possible in calibrating the model.* Different data affect the calibration process differently. Hydraulic-head elevations or changes are among the easiest hydraulic data to obtain outside of rainfall. Hydraulic-head data come in several ways. Most commonly, one obtains depth-to-water data. If an accurate topographic map is available and one is interested in modeling a large area, the relatively crude water-level elevations that may be obtained by subtracting the depth-of-water data from the approximate top of well casing may be useful. However, in modeling small geographical areas, such as often encountered in contaminant transport modeling, very accurate water-level elevations may be needed. In such circumstances, measuring point elevations may be required that are accurate to one one-hundredth of a foot. Calculating accurate groundwater gradients over small horizontal or vertical distances, or when the flow is in very permeable formations where head losses are small, may require water-level elevations at this level of precision. Water-level elevations that represent steady-flow conditions are helpful in calibrating hydraulic conductivity, provided that the net infiltration to the aquifer is known. To understand why one must know the infiltration before being able to calibrate the hydraulic conductivity, or vice versa, consider the steady-state counterpart to equation 1.7: namely,

$$\boldsymbol{\nabla} \cdot \mathbf{K} \cdot \boldsymbol{\nabla} h = Q \tag{1.116}$$

Now assume that the hydraulic conductivity $\mathbf{K}$ in equation 1.116 is homogeneous and isotropic and divide through by this value to give

$$\nabla^2 h = \frac{Q}{K} \tag{1.117}$$

Equation 1.117 reveals that the value of Q and K are confounded in the steady state. Thus you must know one to uniquely define the other.

Now consider the possibility of obtaining time-dependent hydraulic-head information. The most common source of such information is the pumping test, in which a series of observation wells located within the area of influence of a pumping well are measured periodically before, during, and after the discharge of the well has been changed. Under these circumstances, equation 1.7, reproduced below, can be considered in its entirety:

$$\nabla \cdot \mathbf{K} \cdot \nabla h = S_s \frac{\partial h}{\partial t} + Q \tag{1.118}$$

Division by the hydraulic conductivity does not now produce an ambiguity since there are now two terms on the right-hand side of equation 1.118 and therefore two different ratios involving K. Using well-documented techniques (see Batu [18], Walton [19], or the classic Hantush [20]) for a comprehensive discussion of pump-test analysis techniques), one can obtain estimates of both the specific storage S_s and the hydraulic conductivity K, given the pumping well discharge Q. It is also possible to determine the components of the anisotropic form of $\mathbf{K}$ when the pumping test is designed specifically for that purpose. When there are multiple observation wells, a representative value of K can be obtained for each well, providing insight into the heterogeneity of the aquifer. It should be evident that the specific storage cannot be determined from time-invariant hydraulic-head information.

Transient hydraulic-head data can also be measured on a regional scale. In this case, the source of the hydraulic-head variability may be more difficult to determine. Although it is true that the discharge Q in equation 1.118 may include both pumping and precipitation effects, the pump test can be conducted in a manner that isolates the pumping-well influence. In the regional case, the influence of precipitation events and well-pumping variability on measured water levels may be difficult to isolate.

The matter of regional behavior is especially important when automatic parameter estimating programs are employed to assist the analyst in calibrating a model. In such programs, an objective function that contains the differences between observed and computed heads is minimized through appropriate parameter-estimation techniques. In this context, it becomes evident that the more physically different, measurable state variables that are found in the objective function, the more robust will be the parameter estimation process (see Hill [17] for an extensive discussion of the automated calibration process). From this point of view, the discussion under item 2 is particularly germane. In concept, any relevant state variables can be embedded in the objective function, and there are often several possibilities.

4. When calibration appears to be approaching an unacceptably inaccurate asymptote in terms of reproducing the observed state variables, it is prudent to revisit the basic structure of the model. The simplifications inherent in two-dimensional models may violate the fundamental hypotheses inherent in the averaging process required in the reduction of the model from three to two dimensions. Assumptions made about the boundaries of the model may be inappropriate. For example, the assumption of an impermeable boundary at a hydrological divide may be appropriate under steady-state conditions and an absence of nearby pumping, but may not be appropriate when modeling the behavior of a pressure front generated by a change in hydraulic head due to the introduction of a new well or to a change in discharge from an existing one.

   In model calibration the response of the system to incremental changes in parameter values depends on the parameter and its location. Changes in some parameters may induce small changes in state variables, while similar changes in others may induce large changes. Similarly, the impact on the state variables of equal changes in the same parameter located in different parts of the model may elicit quite different state-variable responses. The change in the state variables in response to changes in parameter values is known as sensitivity. Experience with a particular model provides a qualitative sense of the sensitivity of different parameters. An optimization algorithm, on the other hand, computes these values as part of the fitting process and does so automatically. Whether the sensitivities are intuitive or computed as an integral step in an optimization algorithm, their utility in selecting which parameters to change and by what magnitude is conceptually the same.

1.18.2 Model Evaluation Guidelines

1. *Evaluate the model fit.* To assess the accuracy of the model, both objective and subjective measures are helpful. In the subjective category one would consider the visual match between groundwater elevation contours provided by groundwater professionals and those generated from model output by interpolating computer software. One can either examine the two contour plots side by side or digitize the hand-drawn contours and have the computer prepare a contour map of the difference between the two surfaces. Often, a pattern of discrepancies becomes evident. The pattern may show areas of the model where the discrepancies between observed and calculated values are particularly large or where they are especially small. Another variant on this theme is to plot the difference between the observed and calculated values at each observation point. The natural and appropriate tendency is to focus on those areas where the discrepancies are the largest. However, it is inevitable that changes made in one part of the model to improve the state-variable fit will affect not only the area of interest but also calculated values at other parts of the model that were deemed acceptable. The consequence is an iterative, and hopefully convergent, process.

A second commonly used strategy is to calculate the sum-of-squares fit of the calculated and observed values of the state variables. As one improves the accuracy of the model, the normal tendency is to see the sum-of-squares fit improve. In the event that multiple observations over time are available at a well, a sum-of-squares fit to the time series can also be calculated.

2. *Evaluate optimized-parameter values.* Estimated parameter values that are inconsistent with the analysts perception of reality could indicate model error. In other words, having the observed and calculated state-variable match is a necessary but not a sufficient condition for model accuracy. Fundamental errors in the model construction will generate a model that cannot successfully reproduce the observed state variables. To achieve a fit between observed and calculated values in a flawed model may require nonphysical parameter values or values that are inconsistent with information from other lines of evidence. A reality check of all parametric values used in the final version of a model is a must.
3. *Test alternative models.* The optimal model design exhibits at least two attributes: better fit, and more realistic optimal parameter values. If time and resources permit, it is prudent to examine alternative model formulations. The alternative models may be characterized by different boundary conditions, different parameter estimates, different forcing functions, or simply different numerical meshes. A common practice is to reduce the spatial and temporal numerical increments, that is, Δt and Δx_i by a factor of $\frac{1}{2}$, and observe the change in the model solution. If it is perceived to be significant, the procedure should be repeated until the observed change in state variables, for example the hydraulic head, is within acceptable bounds. What is acceptable depends on the importance of simulation accuracy in resolving the issues for which the model was originally developed.

1.18.3 Additional Data-Collection and Model Development Guidelines

Evaluate potential new data. The collection of new information may be cost effective. The need for additional information to enhance the accuracy of the model depends on the trade-off between model accuracy and cost. Models that involve public health and safety directly may demand greater investment for increased accuracy than those required for the investigation of resource exploitation. For example, a model that is designed to predict the movement of a contaminant plume to assess the potential impact of contaminant migration on public health may demand greater forecasting accuracy than a model designed to predict the drawdown generated by a new well installation or the change in pumping rate in an existing well. The trade-off between cost and reduction in optimal design cost for remediation strategies has been investigated quantitatively in the literature. However, in practice, the decision regarding the collection of and utilization of new data in a model is largely determined by the benefits perceived by the client funding the modeling activity.

In making a decision regarding the acquisition of additional data, consideration must be given to the cost of data collection. For example, the cost of additional water-level measurements is generally small relative to the cost of collecting information on field parameters such as hydraulic conductivity. Moreover, some data can be collected quite economically in conjunction with the collection of data that must be obtained for other reasons. For example, measuring the water level in a well prior to sampling it incurs a relatively small additional expense. Improving the accuracy of the water budget in an area of interest, on the other hand, may involve significant additional investment.

1.18.4 Uncertainty-Evaluation Guidelines

The concept of simulation uncertainty has two quite different interpretations. One interpretation addresses the issue of random porous-media properties. In this definition the hydraulic conductivity, for example, is perceived as a random field with known or estimable statistical properties. In the second interpretation, the parameter field is assumed to be deterministic, but the values of the deterministic variables are considered uncertain. The resulting set of stochastic differential equations are similar in their structure. In either case, the mean, variance, and covariance of the parameters must be estimated and the random fields generated, one for each stochastic variable. Once generated, a Monte Carlo type of analysis wherein various realizations of the parameter field are used in simulation can be conducted. In this approach the state-variable field is generated for each parameter field realization. The resulting set of state fields can be used to compute the statistics of the stochastic state field. For problems exhibiting a small variance, a perturbation method can be used to solve the equations directly without Monte Carlo sampling. However, in groundwater simulation, small variances in parameters such as hydraulic conductivity are unusual.

When the approach described above is used to accommodate parameter uncertainty, the issue of calibration has a slightly different connotation. Rather than representing how well the model fits deterministic data, this approach attempts to incorporate the parameter uncertainty into the model itself. Of course the model must still be structurally accurate; that is, the boundary conditions and other structural factors must be correct.

A quite different view on uncertainty comes from an examination of the goodness-of-fit parameters that arise out of either manual or automatic calibration. The final form of the calibrated model will not be exact. Deviations between observed and calculated state variables will exist. In such instances statistical tools such as confidence limits can be used to communicate the degree of discrepancy between the observed and computed states. One can use confidence and prediction intervals to indicate parameter and prediction uncertainty. Hill [16] suggests the following:

> Start by using linear confidence intervals, which can be calculated easily. Test model linearity to determine how accurate these intervals are likely to be. If needed and as possible, calculate nonlinear intervals. . . . Calculate prediction intervals to compare measured values to simulated results. Calculate simultaneous intervals if multiple values are considered or the value is not completely specified before simulation.

Residual or unaccounted for uncertainty in the sense of Hill [16] is a measure of the how well the model represents existing data sets. As in any statistical approach, the projections of state variables are based only on existing data. Any projections that involve stresses that did not exist at the time when the data were collected are not captured in the statistical analysis. The consequence is that the extension of the model statistics to include situations where stresses not represented in prior data sets are to be considered should be treated with care.

1.18.5 Some Rules of Thumb

Calibration of groundwater-flow and transport models is as much an art as a science. Intuition based on experience guides the modeler in making changes to parameters, boundary conditions, or stresses that will provide, with increased iteration, calculated state variables that approach those observed in the field. Nevertheless, there are several rules of thumb presented by Hill [16] that are worthy of mention.

1. "Large-scale deviations suggest adjustments in boundary conditions may be in order." The rationale behind this statement is found in the mathematics of modeling. State variables calculated using a groundwater-flow or transport model constitute the numerical solution to an initial boundary-value problem. If we assume for a moment that the system is homogeneous and has no imposed stresses, such as wells or rainfall, the hydraulic-head solution is determined entirely by the governing equation and the boundary conditions. The addition of heterogeneity and stresses modify this fundamental solution. In fact, in the case of applying stresses, a first approximation to the impact of the stress on the overall solution can be determined by considering the impact of the stress on a homogeneous aquifer with uniform-head boundary conditions and adding the solution to the boundary value problems noted earlier. This is a crude application of the well-known mathematical concept of superposition of linear solutions.

   In summary, large-scale model behavior, especially in the steady state, is governed in large part by the boundary conditions. Consequently, large-scale corrections to the computed state variable, such as hydraulic head, are often achieved by reexamination and informed adjustment to the type or value of boundary conditions.
2. "An inability to match time-dependent changes in regional water levels suggests that forcing functions, hydraulic conductivity or the specific storage may need adjustment." To understand the reasoning behind this observation, it is helpful to rewrite equation 1.7 in a slightly different way:

$$\nabla^2 h = \frac{S_s}{K}\frac{\partial h}{\partial t} + \frac{Q}{K} \qquad (1.119)$$

   where we have assumed for the moment the hydraulic conductivity to be a scalar constant. If we neglect the right-hand side of equation 1.119 the solution for the head h is determined entirely by the boundary conditions. This case is considered in item 1.

Consider now the case where the left-hand side of this equation is neglected. The consequence of this is that the boundary conditions on the problem are irrelevant and there is no gradient in hydraulic head. In this bizarre situation we have an initial-value problem. The solution to this problem depends uniquely on the initial condition $[h(t = 0) = h_0]$ and the two ratios S_s/K and Q/K, that is,

$$\frac{dh}{dt} = -\frac{Q}{K} \cdot \frac{K}{S_s} = -\frac{Q}{S_s} \tag{1.120}$$

Note that since we neglected the space dimension in this simple analysis, the hydraulic conductivity does not appear explicitly in the solution. Thus, to make uniform large-scale changes in time-dependent behavior of the hydraulic head, one should focus on the stresses Q and the specific storage S_s. If these functions are constant, they are also confounded; that is, one can get the same effect whether the numerator is increased or the denominator is decreased.

Let us now complicate the problem a little by considering the commonly encountered problem of flow to a well. Employing radial coordinates, one can write the change in head at a distance r from a well pumping at a rate Q at a time t as (Theis [21])

$$h - h_0 = \frac{Q}{4\pi T} W(u)$$

where $u = r^2 S/4Tt$, $W(u)$ is called the well function, and T is the transmissivity defined as Kl, where l is the aquifer thickness and S is the storage coefficient given by $S = S_s l$. For small values of r or large values of t, one can replace $W(u)$ with the relationship

$$h - h_0 = \frac{2.30Q}{4\pi T} \log \frac{2.25Tt}{r^2 S} = \frac{2.30Q}{4\pi T} \left(\log \frac{2.25}{r^2} + \log \frac{Tt}{S} \right) \tag{1.121}$$

Examination of equation 1.121 reveals that there is a scaling factor, $2.30Q/4\pi T$, that multiplies a log function. Thus the ratio of the well discharge Q divided by the transmissivity T is a scaling factor over all time. Note that since the relationship between Q and T is a ratio, the same scaling effect is realized by either increasing Q or decreasing T.

The function $\log(2.25Tt/r^2S)$ can be rewritten as $\log(2.25/r^2)_+(\log Tt/S)$. The first of these terms is time independent and depends only on the distance from the well r. The second term contains the ratio of the transmissivity T over storage coefficient S. Thus, the effect of this term is to add a time-dependent influence to the change in head, and that influence is amplified by the T/S ratio. Note that the same accelerated head-change behavior can be realized by either increasing T or decreasing S.

Based on this observation, it is evident that if the magnitude of the transient change in head at any time is the issue, one can increase Q or decrease T. If the effect appears to be time dependent in the sense that a deviation is found to increase or decrease over time when compared to measured values, the storage coefficient S is the potential culprit. If, on the other hand, the shape of the hydraulic head surface as one moves away from source or sink areas is the issue, one should consider modifying either Q or T since the ratio Q/T multiplies the term containing the radial distance r from the well.

3. Local head adjustments may be achieved by changing local hydraulic conductivity and leakage values. The rationale for making these changes is based on the concepts discussed in item 2.

In general, final adjustments to flow models require one to visualize how changes in parameters and forcing functions will influence the flow field. This requires that the analyst interact, often extensively, with the model to see what changes in the head field results from specific changes in input.

1.19 PRODUCTION RUNS

Each model should be constructed to answer specific questions. Indeed, the detail and accuracy of a model depends on the questions it is designed to answer. For example, a general understanding of regional water level response to specific pumping strategies may require far less accuracy than that needed for litigation support. In general, one uses models to

- Forecast the flow pattern to be generated in response to remediation design alternatives
- Create flow simulations as part of contaminant transport analyses
- Determine the impact of new water supplies on existent resources

1.20 SUMMARY

The goal of this chapter has been to introduce the concepts of groundwater-flow modeling within the context of the Argus ONE interface and the *PTC* groundwater-flow and transport code. The steps required to formulate and fabricate a model were presented using the Tucson, Arizona, groundwater reservoir as a field example. Information from this chapter provides the groundwater-flow information needed in Part Two to study groundwater-transport modeling.

REFERENCES

[1] Bureau of Reclamation, U.S. Department of the Interior, *Earth Manual*, U.S. Government Printing Office, Washington, DC, 1963.

[2] Anderson, K. E., *Ground Water Handbook*, National Ground Water Association, Westervill, OH, 1991.

[3] Celia, M. A. and G. F. Pinder, "Generalized Alternating-Direction Collocation Methods for Parabolic Equations: I. Spatially Varying Coefficients," *Numerical Methods for Partial Differential Equations*, Vol. 6, No. 3, pp. 193–214, 1990.

[4] Rampe, J. J., "Results of the Tucson Airport Area Remedial Investigation: Phase I. An Evaluation of the Potential Sources of Groundwater Contamination near the Tucson International Airport, Tucson, Arizona," report prepared for Office of Emergency Response and Environmental Analysis, Division of Environmental Health Services, Phoenix, AZ, 1985.

[5] U.S. Environmental Protection Agency, "Ambient Water Quality Criteria for Trichloroethylene," *EPA 44015-80-077*, Office of Water Regulations and Standards, Washington, DC, 1980.

[6] Mock, P. A., B. C. Travers, and C. K. Williams, "Results of the Tucson Airport Area Remedial Investigation: Phase I, Vol. II. Contaminant Transport Modeling," Arizona Department of Water Resources, Hydrology Division, Phoenix, AZ, 1985.

[7] Leake, S. A. and R. T. Hanson, "Distribution and Movement of Trichloroethylene in Ground Water in the Tucson Area, Arizona," *U.S. Geological Survey Water-Resources Investigations Report 86-4313*, 1987.

[8] Drewes, H., "Tectonic Map of Southeastern Arizona," *U.S. Geological Survey-Miscellaneous Map Series Map I-1109*, 1980.

[9] Schmidt, K. D., "Results of the Tucson Airport Area Remedial Investigation: Phase I, Vol. I. Summary Report, Appendices," prepared for Arizona Department of Health Services, Phoenix, AZ, 1985.

[10] Black and Veatch, "Assessment of the Relative Contribution to Groundwater Contamination from Potential Sources in the Tucson Airport Area, Tucson, Arizona," 1987.

[11] Freeze, R. A. and J. A. Cherry, *Groundwater*, Prentice Hall, Englewood Cliffs, NJ, 1979.

[12] Travers, B. C. and P. A. Mock, "Groundwater Modeling Study of the Upper Santa Cruz Basin and Avra Valley in Pima, Pinal and Santa Cruz Counties, Southeastern Arizona," Arizona Department of Water Resources, Phoenix, AZ, 1984.

[13] Hargis and Montgomery, Inc., "Digital Simulation of Contaminant Transport in the Regional Aquifer System, U.S. Air Force Plant No. 44, Tucson, Arizona," 1982.

[14] Anderson, T. W., "Electrical-Analog Analysis of the Hydrologic System, Tucson Basin, Southeastern Arizona," *U.S. Geological Survey Water-Supply Paper 1939-C*, 1972.

[15] Leet, L. D., and S. Judson, *Physical Geology*, Prentice Hall, Englewood Cliffs, NJ, 1959.

[16] Hill, M. C., "Methods and Guidelines for Effective Model Calibration," *Proceedings EWRI, ASCE*, 2000.

[17] Hill, M. C., "Methods and Guidelines for Effective Model Calibration,"*U.S. Geological Survey Water-Resources Investigations Report 98-4005*, 1998.

[18] Batu, V., *Aquifer Hydraulics*, Wiley, New York, 1998.

[19] Walton, W. C., *Practical Aspects of Groundwater Modeling*, National Water Well Association, Westerville, OH, 1984.

[20] Hantush M. S., "Hydraulics of Wells," in *Advances in Hydroscience*, Academic Press, San Diego, CA, 1964, pp. 281–432.

[21] Theis, C. V., "The Relation between the Lowering of Piezometric Surface and the Rate and Duration of Discharge of a Well Using Ground-Water Storage," *Transactions of the American Geophysical Union*, 16th Annual Meeting, Part 2, 1935.

[22] Zhang, Y., "Modeling Three-Dimensional Groundwater Flow in Unsaturated Zone," term paper, Department of Civil and Environmental Engineering, University of Vermont, Burlington, VT, 2001.

[23] Simunek, J. and M. T. van Genucthen, "The HYDRUD-2D Software Package for Simulating the Two-Dimensional Movement of Water, Heat, and Multiple Solutes in Variably-Saturated Media" *U.S. Salinity Laboratory Report*, 1999.

Part 2

Transport Modeling

Groundwater transport generally describes the phenomenon of the movement and evolution of dissolved contaminant species by virtue of groundwater flow and various physical-chemical processes occurring in the subsurface. The major problems of concern are the movement of human-made contaminants and naturally occurring contaminants such as **saltwater in coastal aquifers**.[1]

As in the case of the **flow modeling** described earlier, a **sequence of steps** are normally taken to develop and execute a transport model:

1. Compile water-quality information.
2. Determine the number of physical dimensions.
3. Define the size of the model.
4. Input model boundary conditions.
5. Input model parameters.
6. Input model stresses.
7. Define the model discretization.
8. Run the model.
9. Output concentrations.
10. Calibrate the model.
11. Make production runs.

[1]Saltwater intrusion into coastal aquifers was one of the principal motivations for transport modeling prior to the discovery of groundwater contamination by organic solvents in the early to mid-1970s.

2.1 COMPILATION OF WATER-QUALITY INFORMATION

Water-quality information is very important to the design and calibration of a transport model. Groundwater-quality samples collected from observation wells and, where appropriate, production wells can be used to **calibrate** a model. Sample data can also be used to define the geometry of a contaminant plume and thereby the size of the model needed to forecast contaminant behavior.

Besides groundwater-quality samples, other forms of water-quality information are important. **Soil-gas** sampling involves taking a subsurface sample of air from the unsaturated zone and having it analyzed for volatile compounds. When the concentrations of compounds found in the soil gas is due to groundwater contamination, the soil-gas concentration can often be correlated with the concentrations found in groundwater. Soil gas may also be helpful in locating **surface spills** since contaminants often reside in the **unsaturated zone** in transit to the water table. Residual amounts of contaminants may also remain in the soil until **dissolution** or **volatilization** results in their removal. Soil gas is therefore often useful in providing a preliminary picture of the soil- and groundwater-contaminant distribution in the subsurface. A better design for the location of investigatory wells or soil-sampling borings can be realized from soil-gas information.

Soil samples are often taken in an effort to establish **source areas of contaminants**. Soil samples are most often collected during the drilling process and tested to determine the concentration of contaminants of concern (COCs) in the soil pores and adhering to the grain boundaries. High concentrations of contaminants found at shallow depths in the subsurface generally indicate a nearby source location. High concentrations found in deeper samples can indicate contamination that has been transported by water that has moved upward into the soil by **capillarity** from the water table.

As in the case of the flow model, there are situations where time-dependent water-quality information can be helpful. A plot of the concentration at a well versus time can provide insight into water-quality variability. This, in turn, can be helpful in establishing the adequacy of a calibration. It is reasonable to assume that in the case of a calibrated model, the calculated results will be within the range of concentrations observed at a well.

In summary, water-quality, soil-concentration, and soil-gas concentration information can be used in concert to establish the extent of contamination and in certain instances the source.

Tucson Example

Anecdotal and measured information relevant to groundwater quality is available and can be used in a modeling study of the Tucson Airport area. **Anecdotal information** is often in the form of witness testimony and contained in legal records such as depositions as well as in court testimony. Such information can be very helpful in establishing where contaminant sources exist in the present and where they existed in the past. In some instances information is provided in these records on how individuals disposed of quantities of contaminants at specific locations and over specific time periods. Information on the location of, and mass loading at, contaminant sources is generally used in defining **concentration boundary conditions**. This topic is considered in more detail in Section 2.8.

Concentration information in the form of field measurements is especially valuable in defining the areal extent of the contaminant plume. When this information is plotted on a map, the contaminant plume is more easily distinguished. When the general shape of the plume, and concentration distribution of contaminant in the plume are evident, they may be used in combination with a knowledge of the groundwater flow dynamics to provide insight into the location and concentration levels of contaminant sources. The use of concentration data for defining contaminant sources was discussed in detail in the Tucson example in Section 1.1.

The greatest value of knowledge about the characteristics of the contaminant plume, however, is in **model calibration.** The chemical characteristics of the contaminant plume can provide guidance into the selection and modification of aquifer parameters in the model as they are considered by the analyst in an effort to reproduce, mathematically, the prototype groundwater system. Concentration measurements augment measured water-level elevations as the basic source of information for model calibration. In Section 1.1, we introduced the concept of a **contaminant plume** at Tucson and presented it as Figure 1.2.

At the conclusion of the calibration process, a calibrated mathematical model should produce a close mathematical approximation of the physical processes involved in groundwater flow and transport. The model should provide the information that forms the foundation for characterization of this plume, assuming that its behavior has been simulated for the entire period from the point when contaminants were introduced into the subsurface until the point in time of plume observation.

2.2 PHYSICAL DIMENSIONS

The transport model is subservient to the flow model in terms of dimensionality. If the flow model is three-dimensional, it is most appropriate to also use a three-dimensional transport model. Further, if transport demands a three-dimensional model, the flow model must be three-dimensional also. The transport model depends on the definition of the groundwater flow field as input.

In general, **three-dimensional transport** is warranted whenever there is an aquifer system wherein vertical migration of contamination has occurred or is likely to occur. Whenever the concentrations of the contaminant are high enough to influence the **density** of the groundwater such that the flow is affected, the vertical dimension is essential within the model. This occurs in the case of **saltwater intrusion**, where very high concentrations of salt dissolved in the groundwater increase the density of the groundwater. One should keep in mind that if a vertical dimension is needed and flow to multiple wells is anticipated, a three-dimensional system is required in the transport model.

The vertical integration that was performed for the flow model in Part 1 is applicable to the transport model. However, the assumption of a vertically-uniform concentration in a homogeneous aquifer is less likely to be acceptable than the assumption of a constant vertical head in a similar aquifer. In fact, to the contrary, lack of a vertical gradient may falsely enhance contaminant stratification within the aquifer model.

2.3 MODEL SIZE

The **size of the area to be modeled** depends largely on the questions to be answered by the model. Although it is theoretically possible to have a contaminant plume escape through the boundaries of a flow model, it is generally more advantageous to define a model sufficiently large so as to retain the plume within the model throughout the period of analysis. Since the transport model depends on the definition of the groundwater-flow field as input, the transport model cannot be larger than the groundwater-flow model. However, it may be smaller, since the rate of propagation of pressure which is associated with the flow model is quite different than the rate of movement of dissolved constituents such as contaminants. Unlike the flow model, the contaminant-model boundaries do not have to correspond to physically-meaningful boundaries.

In Figure 2.1 one sees two model boundaries. The larger of these is for flow. Flow depends, in general, more on external boundary conditions that does transport. In this example, the nearest flow boundaries defined by the outermost rectangle are far beyond the anticipated extent of the contaminant plume. Therefore, the contaminant conditions defined on the smaller rectangle illustrated in Figure 2.1 have little to do with the flow boundary conditions defined on the larger rectangle. The main goal of the contaminant boundary conditions is to specify a line beyond which the groundwater is likely to remain uncontaminated. Thus two different domains, and therefore two different boundary-condition-specification locations, are appropriate.

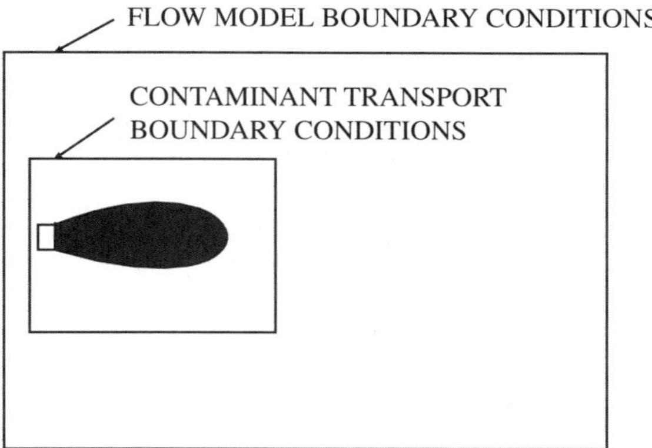

FIGURE 2.1. *Representation of boundary conditions on the flow and transport models.*

2.4 TRANSPORT EQUATION

To understand the rationale behind boundary-condition specification, it is necessary to examine the equation describing contaminant transport. First we write the **species-conservation equation** for flow through a porous medium:

$$[\varepsilon + E(c)]\frac{\partial c}{\partial t} + \nabla \cdot \tilde{\mathbf{q}} + Qc_0 = 0 \tag{2.1}$$

where ε is the porosity, $E(c)$ represents chemical **adsorption** as a function of c, $\tilde{\mathbf{q}}$ is the **mass flux** of the transported species, c is the **concentration** of the species being transported, Q is the **discharge of water** from the model, and c_0 is the **concentration of the transported species in the discharge** Q. First we consider chemical adsorption $E(c)$.

2.4.1 Equilibrium or Adsorption Isotherms

The relationship between the mass of a species adsorbed on the solid phase of a porous medium and that resident in the liquid phase at equilibrium is given by an **adsorption isotherm**. The simplest form of the adsorption isotherm is the **linear adsorption isotherm** given by $I(c) = \alpha_1 c$. The parameter $I(c)$ represents the **mass resident on the solid particles at equilibrium**, c is the initial concentration of the species introduced into a fully saturated system, and α_1 is an experimentally-determined constant. In other words, α_1 is the slope of a line relating the **mass of a given species adsorbed on the solid grains at equilibrium to the initial concentration of the same species in the solution that has been introduced into the system**.

Two nonlinear isotherms include the Freundlich and Langmuir isotherms. The **Freundlich adsorption isotherm** is nonlinear and given by the relationship $I(c) = \alpha_2 c^{\alpha_3}$. The coefficients α_2 and α_3 are, once again, experimentally-determined coefficients. The **Langmuir isotherm** is given by $I(c) = \alpha_4 c/(1 + \alpha_5 c)$ where, as earlier, the coefficients α_4 and α_5 are also determined experimentally.

The rate of change of the concentration of the indicated species due to adsorption is given by

$$E(c)\frac{dc}{dt} = \rho_b \frac{dI(c)}{dt}$$

where ρ_b is the **bulk density of the soil**. Thus for the linear isotherm we have $E(c) = \rho_b \alpha_1$, for the Freundlich isotherm $E(c) = \rho_b \alpha_2 \alpha_3 c^{\alpha_3 - 1}$ or, more simply, $E(c) = \rho_b \alpha c^{\beta}$, and for the Langmuir isotherm we have

$$E(c) = \rho_b \frac{(1 + \alpha_5 c)\alpha_4 {}_{-} \alpha_4 c \alpha_5}{(1 + \alpha_5 c)^2} \quad \text{or} \quad \frac{\rho_b \alpha_4}{(1 + \alpha_5 c)^2}$$

Each of the above is a special case of

$$E(c) = \frac{\rho_b \alpha c^\beta}{(1 + \gamma c)^2} \tag{2.2}$$

where for the linear case $\beta = \gamma = 0$, for the Freundlich isotherm $\gamma = 0$, and for the Langmuir isotherm $\beta = 0$. Once again it should be emphasized that α, β, and γ are experimentally determined coefficients.

If one assumes a linear isotherm, the first term in equation 2.1 becomes

$$[\varepsilon + E(c)]\frac{\partial c}{\partial t} = \varepsilon \left(1 + \frac{\rho_b \alpha}{\varepsilon}\right) \frac{\partial c}{\partial t} \tag{2.3}$$

The coefficient α is often denoted as K_d and called the **distribution coefficient**. The distribution coefficient is considered to be a strong function of the amount of **organic carbon** in the soil. This relationship is described by

$$K_d = K_{\text{oc}} f_{\text{oc}} \tag{2.4}$$

where K_{oc} is the **organic-carbon partitioning coefficient** $[L^3/M]$ and f_{oc} is the **organic-carbon content fraction of the soil** [M/M] (i.e., the number of grams of solid organic carbon per gram of dry soil). While in concept it should be possible to measure K_{oc} by taking the slope of a plot of the K_d versus f_{oc}, this is not the normal protocol that is used to determine K_{oc}.

The parameter K_{oc} is related to the **octonol–water partition coefficient**, K_{ow}. The parameter K_{ow} is a measure of the **hydrophobicity** of organic compounds and is defined as the concentration of a given chemical in octonol divided by the concentration of the same chemical in water. Values of this parameter are readily available in the published literature.

Several relationships exist that define K_{oc} in terms of K_{ow}, depending on the type of organic compound being examined. For example, the relationship

$$K_{\text{oc}} = 0.63 K_{\text{ow}} \tag{2.5}$$

is appropriate for **polynuclear aromatic hydrocarbons** (PAHs). This is a special case of

$$K_{\text{oc}} = a K_{\text{ow}}^b \tag{2.6}$$

Different chemical species will have different coefficients in equation 2.6.

The information appearing as a multiplier of the time derivative in equation 2.3 is often lumped together to give the parameter defined as the **retardation factor**, R_f:

$$R_f \equiv 1 + \frac{\rho_b K_d}{\varepsilon} \tag{2.7}$$

which when substituted into equation 2.3 and subsequently into equation 2.1 yields

$$\varepsilon R_f \frac{\partial c}{\partial t} + \boldsymbol{\nabla} \cdot \tilde{\mathbf{q}} + Qc_0 = 0 \tag{2.8}$$

Retardation is a concept we will revisit shortly.

2.4.2 Mass Flux

The **mass flux** $\tilde{\mathbf{q}}$ is made up of two terms, that is,

$$\tilde{\mathbf{q}} = \mathbf{q}c - \mathbf{D} \cdot \boldsymbol{\nabla} c \tag{2.9}$$

where $\mathbf{q}$ is the **Darcy velocity** (specific discharge) and $\mathbf{D}$ is the **dispersion coefficient**. The first term on the right-hand side of equation 2.9 describes convective mass transport, that is, the movement of the dissolved species by virtue of the average groundwater flow velocity. The second term on the right-hand side of equation 2.9 describes the **dispersive flux**, that is, transport by virtue of the difference between the actual velocity in a given pore and the average velocity $\mathbf{v}$ in a suitably sized control volume (normally, on the order of cubic meters).

The term in equation 2.9 that describes dispersive flux is an assumed **Fickian form for dispersion**. The dispersion coefficient $\mathbf{D}$ is a tensor quantity given as follows:

$$\mathbf{D} = \begin{bmatrix} D_{xx} & D_{xy} & D_{xz} \\ D_{yx} & D_{yy} & D_{yz} \\ D_{zx} & D_{zy} & D_{zz} \end{bmatrix} \tag{2.10}$$

where

$$D_{\alpha\beta} = \alpha_T |q| \delta_{\alpha\beta} + (\alpha_L - \alpha_T) q_\alpha q_\beta / |q| + D_m \delta_{\alpha\beta} \tag{2.11}$$

and α_T is the **transverse dispersivity**, α_L is the **longitudinal dispersivity**, $|q|$ is the magnitude of the Darcy or average pore velocity (i.e., $|q| = \sqrt{q_x^2 + q_y^2 + q_z^2}$), D_m is the **molecular diffusion**, and $\delta_{\alpha\beta}$ is the **Kronecker delta**. The Kronecker delta is defined to be 0 unless $\alpha = \beta$, whereupon it is equal to 1. It is assumed in writing equation 2.11 in terms of two parameters α_L and α_T that the porous medium is isotropic with respect to dispersion.

Let us now introduce equation 2.9 into equation 2.8 to yield

$$\varepsilon R_f \frac{\partial c}{\partial t} + \boldsymbol{\nabla} \cdot (\mathbf{q}c) - \boldsymbol{\nabla} \cdot (\mathbf{D} \cdot \boldsymbol{\nabla} c) + Qc_0 = 0 \tag{2.12}$$

It is computationally convenient to expand the second term in equation 2.12 to give

$$\boldsymbol{\nabla} \cdot (\mathbf{q}c) = \mathbf{q} \cdot \nabla c + c\boldsymbol{\nabla} \cdot \mathbf{q} \tag{2.13}$$

If one ignores the transient behavior of the pressure in the groundwater flow equation, one can see that (see equation 1.5)

$$\nabla \cdot \mathbf{q} = -\,Q \tag{2.14}$$

so that equation 2.13 becomes

$$\nabla \cdot (\mathbf{q}c) = \mathbf{q} \cdot \nabla c - cQ \tag{2.15}$$

Substitution of equation 2.15 into equation 2.12 yields

$$\varepsilon R_f \frac{\partial c}{\partial t} + \mathbf{q} \cdot \nabla c - \nabla \cdot (\mathbf{D} \cdot \nabla c) - Q(c - c_0) = 0 \tag{2.16}$$

It is now clear that the **sink term** (the last term on the left-hand side of equation 2.16) represents the difference in concentration between the fluid leaving the aquifer and the fluid resident in the aquifer. Since these values are normally the same in the case of discharge, this term vanishes in this special case. However, in the case of **recharge**, such as rainfall, where c_0 is equal to zero, or a toxic spill, where c_0 may represent very high values, these two concentrations are normally different. When these values are different the recharge term can cause the resident fluid to either increase or decrease in concentration. When the resident fluid reaches the concentration of the recharged fluid, this term vanishes once again.

Let us now briefly revisit the concept of retardation. To do this we divide equation 2.16 by εR_f, to obtain

$$\frac{\partial c}{\partial t} + \frac{\mathbf{q}}{\varepsilon R_f} \cdot \nabla c - \nabla \cdot \left(\frac{\mathbf{D}}{\varepsilon R_f} \cdot \nabla c \right) - \frac{Q}{\varepsilon R_f}(c - c_0) = 0 \tag{2.17}$$

It is now evident that the effective convective velocity in this equation is not $\mathbf{q}$ but $\mathbf{q}/\varepsilon R_f$. Since R_f is always greater than 1 and $\mathbf{v} = \mathbf{q}/\varepsilon$ is the **pore velocity** (as distinct from the Darcy velocity), it is apparent that retardation tends to slow down the apparent movement of contaminants.

2.4.3 Example of Retardation

Let us assume that we have calculated the Darcy velocity $\mathbf{q}$ and would like to know how long it would take for a contaminant molecule of PAH to travel by convection from point A to point B when the distance between A and B is 100 ft. From equation 2.4 we have

$$K_d = K_{\mathrm{oc}} f_{\mathrm{oc}} \tag{2.18}$$

We also know that for PAHs,

$$K_{\mathrm{oc}} = 0.63 K_{\mathrm{ow}} \tag{2.19}$$

From the literature we find that the value of K_{ow} for PAHs is 200. From equation 2.19 we obtain $K_{oc} = 0.63 \times 200 = 126$ mL/g for PAHs. A measurement made on a field sample of soil indicates that $f_{oc} = 0.02\%$. Thus we have from equation 2.18,

$$K_d = 126 \times 0.0002 = 0.0252 \text{ mL/g}$$

The equation for the retardation coefficient,

$$R_f \equiv 1 + \frac{\rho_b K_d}{\varepsilon} \tag{2.20}$$

requires an estimate of the bulk density and porosity. A field test is used once again to obtain a value of $\rho_b = 1.75$ g/cm^3 and $\varepsilon = 30\%$. We now calculate the retardation as

$$R_f = 1 + \frac{1.75 \text{ g/cm}^3 \times 0.0252 \text{ mL/g}}{0.30} = 1.147 \tag{2.21}$$

If we assume that $\mathbf{q}$ has been determined to be 2 ft/day, we can establish that the particle velocity is

$$\mathbf{v}_p = \frac{2 \text{ ft/day}}{0.30 \times 1.147} = 5.8 \text{ ft/day} \tag{2.22}$$

and the time of travel will be

$$\text{time} = \frac{100 \text{ ft}}{5.8 \text{ ft/day}} = 17.24 \text{ days} \tag{2.23}$$

2.5 CHEMICAL REACTIONS

Chemical reactions, especially **biochemical reactions**, can have a significant impact on the evolution of a contaminant plume consisting of organic compounds. The role of chemical reactions in groundwater contamination hydrology is a rapidly evolving field, and entire monographs have been dedicated to this topic (see, e.g., Bedient et al. [2]). In this section we consider chemical evolution at a level of sophistication consistent with current field practice.

As a point of departure, we will consider equation 2.16 modified to include a reaction term R_c; that is,

$$\varepsilon R_f \frac{\partial c}{\partial t} + \mathbf{q} \cdot \nabla c - \nabla \cdot (\mathbf{D} \cdot \nabla c) - Q(c - c_0) + R_c = 0 \tag{2.24}$$

where R_c is the rate of decay of the species c. We will assume that the compound c is degraded according to the **first-order rate equation**,

$$R_c = \frac{dc}{dt} = -\lambda c \tag{2.25}$$

where λ is the rate of reaction ($-\lambda$ is the rate of decrease) and c is defined in terms of number of molecules of the chemical. Since c is typically defined in terms of mass per unit volume of solution in most practical applications, the number of molecules can be determined by dividing by the molecular weight. One can show that λ is related to the **half-life** $t_{1/2}$ of the chemical through the relationship

$$\lambda = \frac{\ln 2}{t_{1/2}} = \frac{0.693}{t_{1/2}} \tag{2.26}$$

The half-life $t_{1/2}$ is the time required for the number of molecules of compound c, say N_c, to decay to $N_c/2$.

Equation 2.25 is often used to describe the decay of a radioactive compound and is also used to describe the biodegradation of an organic compound. In the case of a radioactive compound, **daughter products** are created from the decay of the compound. For example,

$$_{92}\mathrm{U}^{238} \rightarrow {}_{90}\mathrm{Th}^{234} + \alpha \text{ particle} \tag{2.27}$$

$$_{90}\mathrm{Th}^{234} \rightarrow {}_{91}\mathrm{Pa}^{234} + \beta \text{ particle} \tag{2.28}$$

where the subscript is the **atomic number** (number of protons) and the superscript is the **mass number** (number of protons plus neutrons). Gamma radiation accompanies these transformations but does not change the atomic number or atomic weight.

If we assume that compound c_1 decays to produce isotope c_2, which in turn decays to produce isotope c_3, the following set of equations results:

$$\varepsilon R_f \frac{\partial c_1}{\partial t} + \mathbf{q} \cdot \nabla c_1 - \nabla \cdot (\mathbf{D} \cdot \nabla c_1) - Q(c_1 - c_{0_1}) + \lambda_1 c_1 = 0 \tag{2.29}$$

$$\varepsilon R_f \frac{\partial c_2}{\partial t} + \mathbf{q} \cdot \nabla c_2 - \nabla \cdot (\mathbf{D} \cdot \nabla c_2) - Q(c_2 - c_{0_2}) - \lambda_1 c_1 + \lambda_2 c_2 = 0 \tag{2.30}$$

$$\varepsilon R_f \frac{\partial c_3}{\partial t} + \mathbf{q} \cdot \nabla c_3 - \nabla \cdot (\mathbf{D} \cdot \nabla c_3) - Q(c_3 - c_{0_3}) - \lambda_2 c_2 + \lambda_3 c_3 = 0 \tag{2.31}$$

The decay rate for various isotopes varies enormously. For example, the half-life of $_{92}\mathrm{U}^{238}$ is 4.5×10^9 years, while that for $_{90}\mathrm{Th}^{234}$ is 24.5 days.

Let us now turn our attention to the case of biochemical degradation of organic compounds typically found in groundwater. As an example we consider the transformation of TCE through biochemical degradation. Under suitable environmental conditions, TCE degrades to 1,1 DCE, *trans*-1,2-DCE, and *cis*-1,2-DCE. The biodegradation rate of TCE to *cis*-1,2-DCE is tabulated either in terms of the rate constant or half-life (see, e.g., Bedient et al. [2] or Lorah et al. [3]). Adopting, as an example, the TCE biodegradation rate constant of 0.31 per day [3], one can determine the half-life using equation 2.26, that is,

$$\lambda = \frac{\ln 2}{t_{1/2}} = \frac{0.693}{t_{1/2}}$$

or

$$t_{1/2} = \frac{0.693}{\lambda} = \frac{0.693}{0.31} = 2.22$$

Equations of the form found in equations 2.29–2.31 can now be used to determine the transport of TCE and *cis*-1,2-DCE. One equation will be required for the host compound and one for each daughter product.

2.6 MODEL BOUNDARY CONDITIONS

Given the second-order derivatives appearing in equation 2.16, one has the option of specifying either the concentration value (first-type condition) or its derivative (second-type condition) as a boundary condition. One can also specify a combination of the concentration and its derivative as a boundary condition (third-type condition). While the specification of a concentration value is quite straightforward, the matter of specifying a mass flux is more subtle. To understand this, we return once again to the definition of the flux, that is,

$$\tilde{\mathbf{q}} = \mathbf{q}c - \mathbf{D} \cdot \nabla c \tag{2.32}$$

If we wish to allow contaminants to move across a boundary by convection only, the appropriate boundary condition is a **specified dispersive flux of zero**. In this case we require that $\partial c/\partial \mathbf{n} = 0$ along the boundary perimeter. The flux across the boundary then becomes

$$\tilde{\mathbf{q}} = \mathbf{q}c \tag{2.33}$$

Dissolved species still move across the boundary, but only via convection. This is the most commonly used boundary condition on the perimeter of a contaminant-transport model.

If we want to have no flux across the perimeter boundary (i.e., neither convective nor dispersive), the condition would be

$$\mathbf{q}c - \mathbf{D} \cdot \nabla c = 0 \tag{2.34}$$

This is a form of the Robbins or third-type boundary condition. It is seldom used in practice because if the flow across the boundary is zero, q_n will be zero and the convective flux is automatically eliminated. Under such circumstances, this boundary condition becomes a second-type condition, that is,

$$\mathbf{D} \cdot \nabla c = 0 \tag{2.35}$$

Of course, if a nonzero total flux q_n is required, equation 2.32 is the appropriate expression.

Although unusual in the case of groundwater flow, it is quite appropriate in the case of groundwater transport modeling to place a series of constant concentration

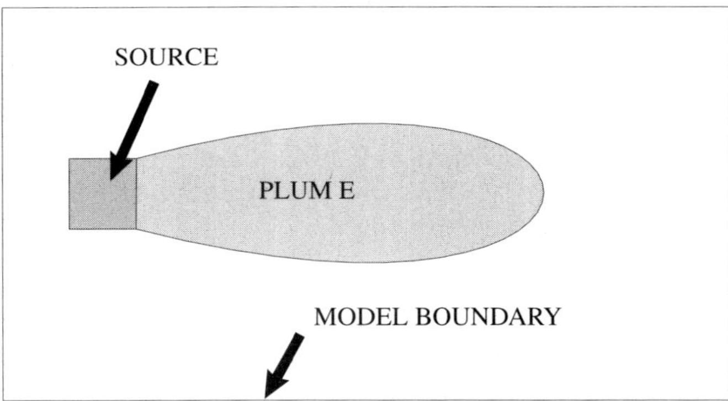

FIGURE 2.2. *Diagrammatic representation of internal concentration boundary conditions.*

nodes in the interior of a model. Consider, for example, the case presented in Figure 2.2.

The outer rectangle in this figure is meant to depict a concentration boundary condition on the perimeter of the model. The internal shaded rectangle denotes a concentration boundary condition that is internal to the model perimeter. This kind of internal boundary condition is used to denote a **contaminant source**. The internal boundary condition is typically either a specified concentration (Dirichlet or first-type) boundary condition or a specified flux (Neumann or second-type) boundary condition. These boundary conditions can be a function of time when such a functional relationship is known.

The question naturally arises as to the best strategy for handling concentration boundary conditions. The answer is that specified concentrations are generally easier to handle computationally than specified fluxes when they can be established. The reason for this is that when a specified flux is prescribed, one must be sensitive to the fact that mass conservation must be maintained, thus the model must adjust to whatever mass flux is introduced. If the estimated flux is inaccurate, bizarre concentration values can result from the model's efforts to compensate numerically for the inaccurate flux. Unlike when the concentration boundary condition is specified, there is no naturally occurring mathematical constraint as to the concentration level that may be generated at the nodes where a specified flux is indicated.

Tucson Example

Run-Time Configuration While groundwater-transport simulation builds on the information we provided for the groundwater-flow calculations, some additional information is required. Let us begin by reexamining the *PTC Configuration window*. We access this window by first clicking on *PIEs* on the menu bar, followed by selecting *Edit Project Info*. . . . The resulting screen configuration is shown in Figure 2.3. The *General* tab dialog box is provided.

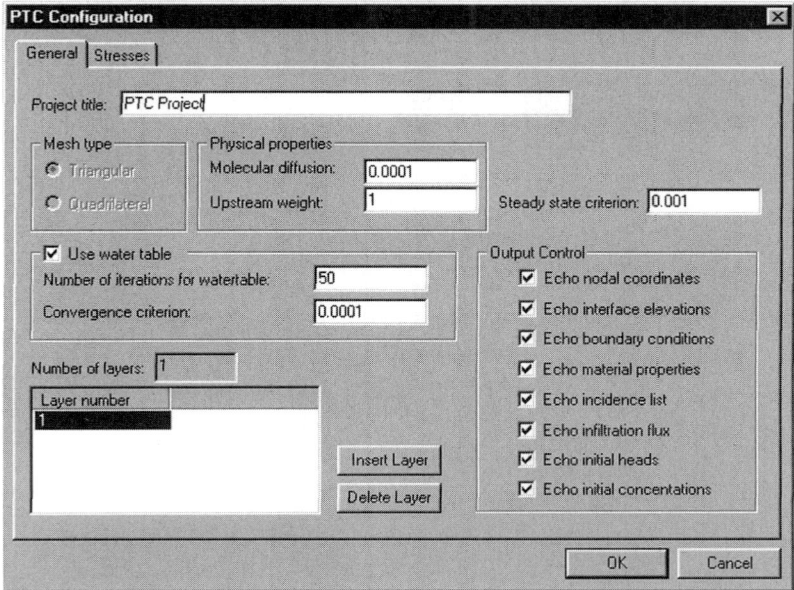

FIGURE 2.3. *The PTC Configuration dialog box where information on molecular diffusion and upstream weighting must be added to accommodate species transport.*

As noted in Chapter 1, selection of the *Use water table* option is associated with the flow calculations and informs *PTC* of the existence (turn on the check box) or nonexistence (turn off the check box) of a water table in the prototype system. The *Number of iterations for watertable* and *Convergence criterion* text boxes must be completed to assure that the nonlinear solution to the water-table calculation is accurate.

The *Steady state criterion* can be an important parameter in transport. It defines the cutoff change in water-table elevation that is considered to be indicative of steady-state flow. Once the cutoff change is realized, the transient part of the flow equation is no longer activated for the stress period specified. The goal of this strategy is to minimize computational effort since the transient behavior of flow is no longer of importance once steady state is reached.

Two new pieces of information are required in this dialog box. In the *Molecular diffusion* text box the value of molecular diffusion for the species of concern must be obtained from standard tables and entered. The value of the upstream-weighting coefficient is placed in the *Upstream weight* text box. The upstream-weighting coefficient has a value between zero and 1. A value of zero provides no upstream-weighting and a value of 1 provides the maximum upstream weighting effect. In Section 2.7 the concept of upstream weighting is presented in detail and we defer further consideration of it until that point in the text.

Let us now activate the *Stresses* tab of the *PTC Configuration* dialog box. The resulting window is provided as Figure 2.4. To perform transport calculations we must

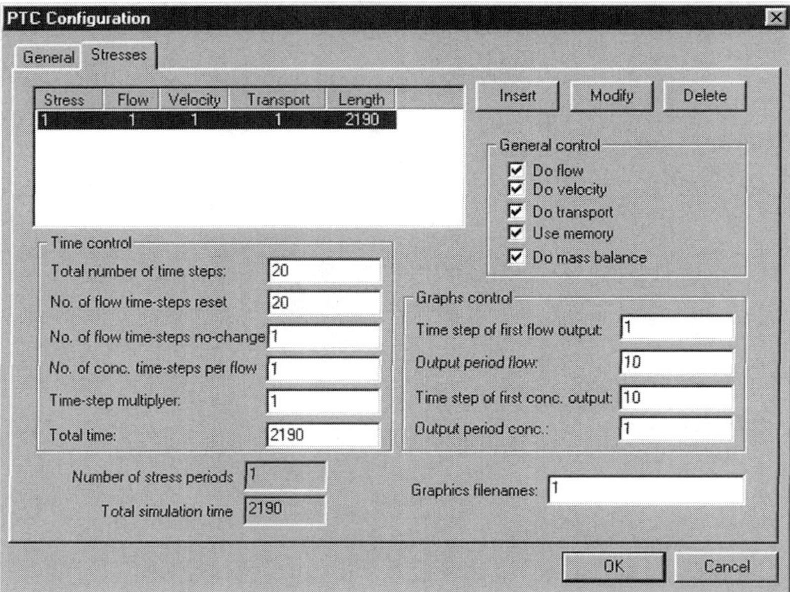

FIGURE 2.4. *PTC Configuration window wherein the parameters have been set for transport simulation. Note that the velocity and the transport options have been selected.*

activate the *Do velocity* and *Do transport* options. The *Do mass balance* option may be activated if a mass-balance error calculation is needed. One should keep in mind, however, that the relationship between the mass-balance error and the numerical error are, in general, quite different. One can have an accurate mass balance and an inaccurate simulation, and vice versa, depending on the numerical method used.

In the series of text boxes associated with time control, several items require attention. Although much of the information in this dialog box was discussed in Chapter 1, we revisit it at this point. The *Total time* now reflects the simulation period for transport and concomitantly flow. The *Total number of time steps* represents the number of time steps required to accommodate the *Total time*. The *Time step multiplyer* box is used to record the value by which each existing flow time step is multiplied in determining the length of the new time step. For example, if the size of the existing time step for time step n is Δt_n, then with a multiplier of 1.5, the new time step size for time step $n+1$ would be $\Delta t_{n+1} = 1.5 \times \Delta t_n$. The *No. of flow time steps reset* text box entry is where one specifies the number of time steps for the flow calculation, after which the time step will be reset using the time-step multiplier. If the value in the *No. of flow time-steps reset* box is given as 5, the first modified time step will be the sixth, and it will be used for steps six through ten. The *No. of time-steps no-change* records the total number of time steps after which the flow time step is no longer changed by the time-step multiplier. In our previous example, if the value placed in this text box was set to 6, the time-step multiplier would be activated only once.

The *No. of conc. time-steps per flow* text box value indicates the number of times the transport equations will be solved for each flow time step. In other words, if the value in this box is set to 3, then given the flow time step is Δt_n^f, the time step for transport Δt_n^t would be $\Delta t_n^t = \Delta t_n^f/3$. The rationale behind this option is the recognition that time steps suitable for flow may be too large to provide an accurate transport calculation. This concern is particularly relevant at long times when a time-step multiplier greater than one is used for flow. The reason for this lies in the physics underlying the transport simulation. While the rate of change in head may be markedly different over the course of a simulation, changes in concentration tend to be similar throughout the simulation, due to the fact that the flow velocity, which dictates the dominating phenomenon of convection, does not generally change markedly after a short initial period.

In considering the matter of time-step size it is helpful to realize that the flow equations do not need to be solved at each transport step because the pressure response is very rapid relative to the transport response. The result is obvious computational savings. Second, since the concentration time step is held constant during the transport calculations made during the period of time that flow is not being updated, the coefficient matrices involved in the transport calculation do not change. Due to the invariant velocity field and the constant time step, the matrices in the matrix equation do not change at each transport time step. As long as the matrix equation does not change, the transport equations do not need to be solved because a back-substitution step using the upper triangular form of the coefficient matrices is all that is needed to calculate the transport. This requires very little effort relative to solving the matrix equation since most of the effort associated with solving a matrix equation is devoted to upper triangularization of the coefficient matrices. The rationale behind the efficiencies defined above will become more evident after reading Section 2.7.

The *Graphs control* group of text boxes for transport allow one to specify different graphical output strategies for transport than for flow. Generally speaking the number of plots required to visualize transport behavior is larger than for flow.

Initial Conditions The distribution of contaminants in the aquifer at the initiation of the simulation period must be specified as **initial conditions**. When the simulation period corresponds to the period during which the contaminant plume evolved (i.e., the beginning of the simulation is the same point in time that the contaminants entered the ground), the initial condition is easily accommodated. In such circumstances the concentration everywhere in the aquifer will be zero, or background. Only the source locations will be nonzero and this circumstance is accommodated via the boundary conditions, as discussed in Section 2.8.

To input initial conditions for concentration, go to the *Layers* window and make the *Initial Concentrations L1* layer active (Figure 2.5). If the goal is to introduce a uniform initial concentration, click *View* from the menu bar and then *Layers...* from the options. Next, click f_x and type in the desired value (generally, this is zero for an uncontaminated aquifer). The resulting *Expression* dialog box allows for simple arithmetic calculations to assist in providing input values as well as more sophisticated options that are described in the Argus ONE manual. If a variable initial

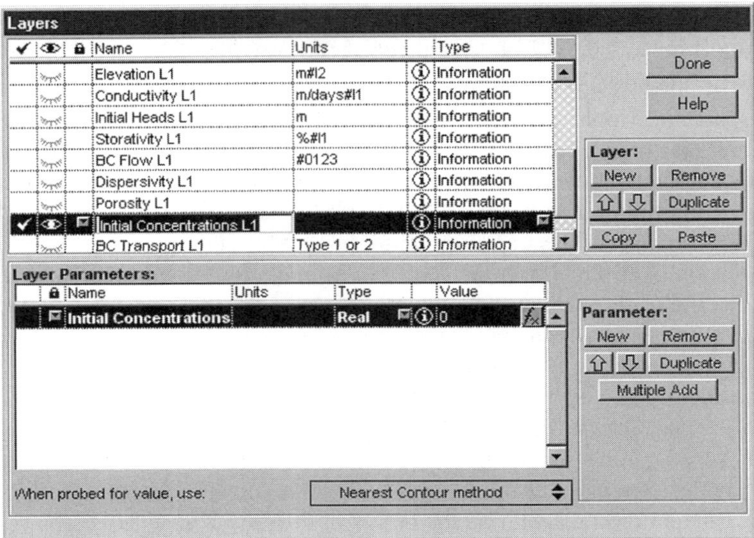

FIGURE 2.5. *Use of the Layers window to specify initial concentration values.*

concentration is desired, the various input options described in the Tucson examples in Sections 1.4 and 1.13 are applicable.

Boundary Conditions Concentration **boundary conditions** are input in a manner analogous to flow boundary conditions as described in Section 1.3. Along the perimeter of the model domain, one of the three standard types of boundary conditions must be specified. If no specification is made, **a zero-dispersive flux condition is assumed as default**.

To define the perimeter boundary conditions, make the *BC Transport L1* layer active, as shown in Figure 2.6. Using the contour tool, define a polygon around the segment of boundary of interest. Be careful to keep the polygon tight to the boundary to avoid specification of unintended nodes as boundary condition nodes. Having defined the segment of interest, double-click on any part of the area enclosed by the polygon. The double click will reveal the *Contour Information* window, as shown in Figure 2.6. In this window specify the type (1 for Dirichlet or known concentration and 2 for a Neumann or specified dispersive-concentration flux) and the value of the boundary condition. Repeat this action for each segment of the boundary perimeter.

As noted above an often-encountered form of the concentration boundary condition is associated with contaminant sources located on the interior of the model. You may recall that in the Tucson example in Section 1.11 we described a procedure for defining point boundaries. The same method is used here to specify constant concentration boundary conditions defined at a source location. The locations identified earlier as concentration sources in the *PTC Domain Outline* layer are now revisited by first selecting this layer. Copy only the points identified with contamination

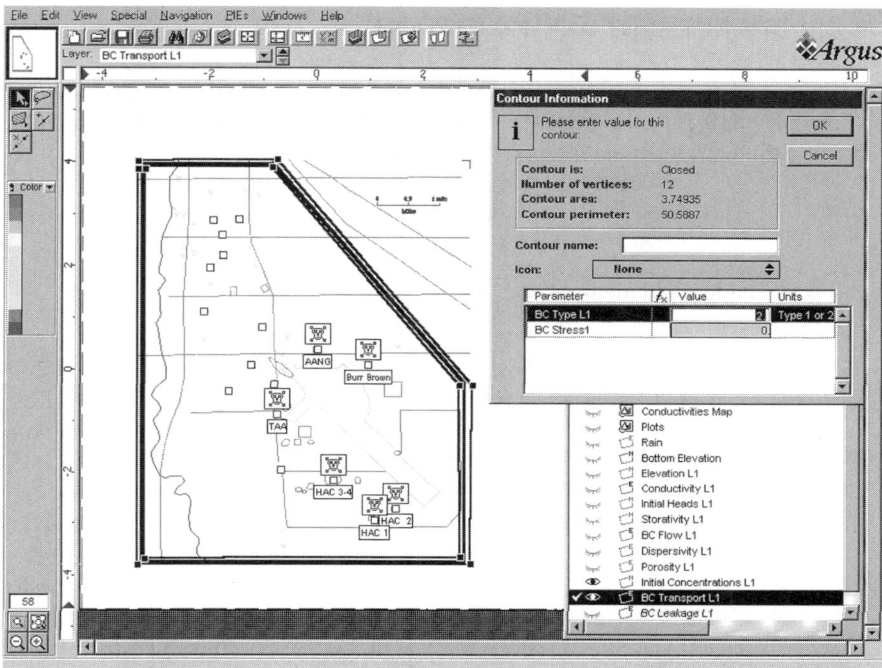

FIGURE 2.6. The BCcTypL1 layer used to define concentration boundary conditions on the perimeter of the model.

sources in the domain. If any contour is already selected, unselect all by clicking anywhere outside the domain boundary. Click on the first point contour. To add a second contour to the selection, press and hold the Shift key and click on the next point contour (the contours selected are shown as a black point on the screen). With the Shift key pressed, you add contours to the selection one by one by clicking on them. When finished, select *Edit* | *Copy* from the menu. Activate the *BC Transport L1* layer and, using the *Paste* command from the *Edit* menu, paste the information into this layer. By clicking on the source location desired, a *Contour Information* window similar to that shown in Figure 2.6 appears. The information that must be provided is the type of boundary to be specified and its magnitude.

The choice of point first-type boundary conditions used in the Tucson model can be examined by activating the *BC Transport L1* layer and then double-clicking on the skull-and-crossbones symbols used to identify the various concentration sources. One will observe that a variety of concentration values have been used at various locations in the model, depending on the chemical characteristics of the different sources. The selection of these first-type boundary condition concentrations is based on a review of the material available in the various documents that describe the contaminant hydrology of the site.

2.7 FINITE-ELEMENT APPROXIMATION

We begin our discussion of the finite-element approximation to the transport equation by first considering the finite-element approximation to equation 2.16 (further details regarding the finite-element method can be found in Section 1.11):

$$\varepsilon R_f \frac{\partial c}{\partial t} + \mathbf{q} \cdot \nabla c - \nabla \cdot (\mathbf{D} \cdot \nabla c) - Q(c - c_0) = 0 \tag{2.36}$$

Next, we provide a statement of the approximation for $c(\mathbf{x}, t)$:

$$c(\mathbf{x}, t) \simeq \hat{c}(\mathbf{x}, t) = \sum_j^J c(z, t)_j \phi(x, y)_j \tag{2.37}$$

Next, we write the **Galerkin approximation** to the transport equation:

$$\int_{\Omega_{xy}} \left[\varepsilon R_f \frac{\partial \hat{c}}{\partial t} + \mathbf{q} \cdot \nabla \hat{c} - \nabla \cdot (\mathbf{D} \cdot \nabla \hat{c}) - Q(\hat{c} - c_0) \right] \phi(x, y)_i \, d\Omega = 0,$$
$$i = 1, \ldots, I \tag{2.38}$$

where Ω_{xy} is the model domain projected onto the (x, y) coordinate system and I is the number of nodes in the model.

Application of **Green's theorem** to the second-order term defined in (x, y) yields

$$\begin{aligned} &\int_{\Omega_{xy}} \left[\varepsilon R_f \frac{\partial \hat{c}}{\partial t} + \mathbf{q} \cdot \nabla \hat{c} - Q(\hat{c} - c_0) \right] \phi(x, y)_i \, d\Omega \\ &+ \int_{\Omega_{xy}} (\mathbf{D} \cdot \nabla_{xy} \hat{c}) \cdot \nabla_{xy} \phi(x, y)_i \, d\Omega - \int_{\Omega_{xy}} D_{zz} \frac{\partial^2 \hat{c}}{\partial z^2} \phi(x, y)_i \, d\Omega \\ &- \int_{\partial\Omega} \mathbf{D} \cdot \nabla_{xy} \hat{c} \phi(x, y)_i \cdot \mathbf{n} \, da = 0, \qquad i = 1, \ldots, I \end{aligned} \tag{2.39}$$

where $\nabla_{xy}(\cdot)$ is the gradient operator defined in the areal, or horizontal plane. Note that we have not used Green's theorem in the z-coordinate direction.

Substitution of equation 2.37 into equation 2.39 results in

$$\begin{aligned} &\int_{\Omega_{xy}} \left[\varepsilon R_f \frac{\partial \sum_j^J c(z, t)_j \phi(x, y)_j}{\partial t} \right] \phi(x, y)_i \, d\Omega \\ &+ \int_{\Omega_{xy}} \left[\mathbf{q}_{xy} \cdot \nabla_{xy} \sum_j^J c(z, t)_j \phi(x, y)_j + q_z \frac{\partial}{\partial z} \sum_j^J c(z, t)_j \phi(x, y)_j \right] \phi(x, y)_i \, d\Omega \\ &- \int_{\Omega_{xy}} Q \left[\sum_j^J c(z, t)_j \phi(x, y)_j - c_0 \right] \phi(x, y)_i \, d\Omega \end{aligned}$$

$$+\int_{\Omega_{xy}} \left[\mathbf{D} \cdot \boldsymbol{\nabla}_{xy} \sum_j^J c(z,t)_j \phi(x,y)_j\right] \cdot \boldsymbol{\nabla}_{xy}\phi(x,y)_i \, d\Omega$$

$$-\int_{\Omega_{xy}} D_{zz}\frac{\partial^2}{\partial z^2}\sum_j^J c(z,t)_j\phi(x,y)_j\phi(x,y)_i \, d\Omega$$

$$-\int_{\partial\Omega} \mathbf{D} \cdot \boldsymbol{\nabla}_{xy}\hat{c}\phi(x,y)_i \cdot \mathbf{n} \, da = 0, \qquad i = 1, \ldots, I \tag{2.40}$$

The reason that we have not substituted for $\hat{c}$ in the surface integral is that this term is either known via a second-type boundary condition, or in the case of a first-type boundary condition is not required because no equation is written at the node where the concentration is specified.

We can now rearrange this equation to obtain

$$\sum_j^J \frac{\partial c(z,t)_j}{\partial t}\int_{\Omega_{xy}} \varepsilon R_f \phi(x,y)_j\phi(x,y)_i \, d\Omega$$

$$+\sum_j^J c(z,t)_j \int_{\Omega_{xy}} \mathbf{q}_{xy} \cdot \boldsymbol{\nabla}_{xy}\phi(x,y)_j\phi(x,y)_i \, d\Omega$$

$$+q_z\frac{\partial}{\partial z}\sum_j^J c(z,t)_j\int_{\Omega_{xy}} \phi(x,y)_j\phi(x,y)_i \, d\Omega$$

$$-\int_{\Omega_{xy}} Q\left[\sum_j^J c(z,t)_j\phi(x,y)_j - c_0\right]\phi(x,y)_i \, d\Omega$$

$$+\sum_j^J c(z,t)_j\int_{\Omega_{xy}} [\mathbf{D} \cdot \boldsymbol{\nabla}_{xy}\phi(x,y)_j] \cdot \boldsymbol{\nabla}_{xy}\phi(x,y)_i \, d\Omega$$

$$-\frac{\partial^2}{\partial z^2}\sum_j^J c(z,t)_j\int_{\Omega_{xy}} D_{zz}\phi(x,y)_j\phi(x,y)_i \, d\Omega$$

$$-\int_{\partial\Omega} \mathbf{D} \cdot \boldsymbol{\nabla}_{xy}\hat{c}\phi(x,y)_i \cdot \mathbf{n} \, da = 0, \qquad i = 1, \ldots, I \tag{2.41}$$

Compare equation 2.41 with its counterpart from Section 1.11:

$$\int_{\Omega_{xy}} \left\{-\mathbf{K} \cdot \boldsymbol{\nabla}_{xy}\left[\sum_{j=1}^J h(z,t)_j\phi(x,y)_j\right]\right\} \cdot \boldsymbol{\nabla}_{xy}\phi(x,y)_i \, d\Omega$$

$$+\int_{\Omega_{xy}} \frac{\partial}{\partial z}\left[K_{zz}\frac{\partial}{\partial z}\sum_{j=1}^J h(z,t)_j\right]\phi(x,y)_j\phi(x,y)_i \, d\Omega$$

$$-\int_{\Omega_{xy}} \left\{ S \frac{\partial \left[\sum_{j=1}^{J} h(z,t)_j \phi(x,y)_j \right]}{\partial t} + Q \right\} \phi(x,y)_i \, d\Omega$$

$$+ \int_{\partial\Omega_{xy}} \mathbf{K} \cdot \frac{\partial \hat{h}}{\partial \mathbf{n}} \phi(x,y)_i \, dl, = 0, \qquad i = 1, \ldots, I \tag{2.42}$$

One observes that these equations are essentially the same in form except for the term describing convection:

$$\sum_j^J c(z,t)_j \int_{\Omega_{xy}} \mathbf{q}_{xy} \cdot \boldsymbol{\nabla}_{xy} \phi(x,y)_j \phi(x,y)_i \, d\Omega$$

$$+ q_z \frac{\partial}{\partial z} \sum_j^J c(z,t)_j \int_{\Omega_{xy}} \phi(x,y)_j \phi(x,y)_i \, d\Omega \tag{2.43}$$

Hereinafter we focus on these two new terms since we have previously dealt with terms of the same form as the others.

A finite-difference approximation is needed for the z derivative appearing in equation 2.43. This we write as

$$q_z \frac{\partial c}{\partial z}\bigg|_i = q_{zB} \beta \frac{c_{i+1,n+1} - c_{i,n+1}}{\Delta z_B} + q_{zA}(1-\beta) \frac{c_{i,n+1} - c_{i-1,n+1}}{\Delta z_A} \tag{2.44}$$

where the locations A and B are defined in Figure 1.43 and $\beta \geq 0.5$ is the upstream-weighting parameter.

Although the formulation above appears straightforward, and indeed it is from a purely numerical-approximation point of view, in practice the approximation of the derivative in the convective term gives rise to serious numerical challenges when it is large relative to the dispersive term in equation 2.41. To see this, let us consider the simplified, one-dimensional form of equation 2.36:

$$\varepsilon R_f \frac{\partial c}{\partial t} + q \frac{\partial c}{\partial x} - D \frac{\partial^2 c}{\partial x^2} = 0 \tag{2.45}$$

where q and D are now scalar quantities. A proportionality constant that is indicative of the magnitude of convection when compared to dispersion is given by

$$P_e = \frac{q \, \Delta x}{D} \tag{2.46}$$

where P_e is defined as the **Peclet number**.

Generally speaking, to assure that the model yields an oscillation-free concentration profile, this ratio should not be allowed to exceed 2. Such a constraint results in an upper limit on the size of the space step Δx. Although this limitation has nothing to do with numerical stability in the model, it has a lot to do with the accuracy of the

concentration can be written (holding time constant) as

$$c_{i+1,n+1} = c_{i,n+1} + \left.\frac{\partial c}{\partial x}\right|_{i,n+1} \Delta x + \frac{1}{2!}\left.\frac{\partial^2 c}{\partial x^2}\right|_{i,n+1} \Delta x^2 + O(\Delta x^3) \tag{2.49}$$

If one now writes a similar expression to establish the concentration at $c_{i,n+1}$ in terms of information at $c_{i-1,n+1}$, it is possible to subtract the resulting two equations to obtain a representation of the convective term at x_i that has no second-order term. However, if any point other than point x_i is used as the reference location for the first derivative evaluation (i.e., the point on which one is standing when writing the approximation), the second-order term does not vanish. Under these conditions, the approximation for the first-order term will have a second-order contribution. By ignoring the second-order contribution, artificial diffusion (dispersion) is added to the numerical model. This artificial diffusion tends to smear a sharp concentration front and suppress any oscillations that might otherwise occur.

To exploit the use of artificial diffusion, one can use an **upstream-weighted** formulation in approximating the convective term, such as (from equation 2.49)

$$\left.\frac{\partial c}{\partial x}\right|_{i,n+1} = \frac{1}{\Delta x}\left[c_{i+1,n+1} - c_{i,n+1} - \frac{1}{2!}\left.\frac{\partial^2 c}{\partial x^2}\right|_{i,n+1} \Delta x^2 + O(\Delta x^3)\right] \tag{2.50}$$

We can see from equation 2.50 that inherent in the representation of the first derivative in concentration is a second derivative term. Ignoring this term adds numerical diffusion (dispersion) to the equation.

A similar analysis can be generated for the finite-element equations, although the analysis is more complex. In essence, one uses one-half of the standard **chapeau function** to represent an upstream weighted approximation (Figure 2.9).

In this kind of formulation, only the portion represented by a-b-c is used as the weighting function for the first-derivative term. This is no longer a Galerkin formulation but a generalization that will allow different terms in the equation to be weighted by different functions. It is also possible to use a variant on the finite-

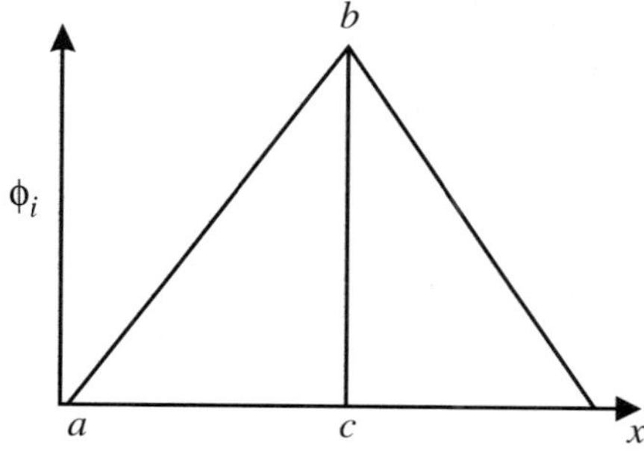

FIGURE 2.9. *Diagrammatic representation of upstream weighting using chapeau basis functions.*

element approach above that will permit the use of variable upstream-weighting. This is described in Huyakorn and Pinder [1] and implemented in *PTC*.

Before we leave the finite-element discussion, there is one more interesting point. In equation 2.39 we applied the Green's theorem to the second-order term. But we could have also applied it to the first-order spatial term. In other words, we could have done the following (see equation 2.12):

$$\int_{\Omega}[\nabla \cdot (\mathbf{q}\hat{c}) - \nabla \cdot (\mathbf{D} \cdot \nabla \hat{c})]\phi_i \, d\Omega = -\int_{\Omega}(\mathbf{q}\hat{c} - \mathbf{D} \cdot \nabla \hat{c}) \cdot \nabla \phi_i \, d\Omega + \int_{\partial\Omega}(\mathbf{q}\hat{c} - \mathbf{D} \cdot \nabla \hat{c}) \cdot \mathbf{n} \, da \qquad (2.51)$$

At first blush this looks like a handy formulation, since one automatically obtains a third-type boundary condition in the process. Based on experience, however, we have found that this formulation, in general, does not work very well. To understand why, consider writing the boundary term for each finite element; that is, for element e we have

$$I_e = \int_{\partial\Omega_e}(\mathbf{q}\hat{c} - \mathbf{D} \cdot \nabla \hat{c}) \cdot \mathbf{n} \, da_e \qquad (2.52)$$

Consider the two elements illustrated in Figure 2.10. Along line A–B, the integral appearing in equation 2.52 that represents the boundary term associated with element 1 must cancel with the same term associated with element 2. However, the derivative of the head in each element is a different constant and, therefore, also is the value of $\mathbf{q}$. Moreover, the concentration along line A–B is the same whether viewed from element 1 or element 2. Thus it is impossible for the term

$$\int_{\partial\Omega_e}(\mathbf{q}\hat{c}) \cdot \mathbf{n} \, da_e \qquad (2.53)$$

to cancel. It has different values when viewed from each element. It appears that by not accounting specifically for this change in mass, an error of significant proportions is generated. Whether the inclusion of this term (equation 2.53) in the approximating equation would rectify the situation remains an open question.

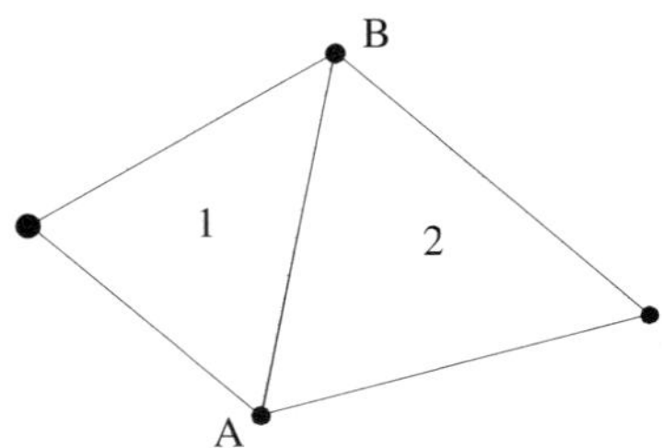

FIGURE 2.10. *Definition sketch for discussion of mass balance on boundary.*

2.8 BOUNDARY CONDITIONS

In defining the boundary conditions for this equation, one proceeds as in the case of the flow equation.

2.8.1 First-Type Boundary Condition

For a constant-concentration boundary condition, the concentration value at the boundary node is set equal to the known value; that is, $c_{i,n} = c_{i0}$.

2.8.2 Second-Type Boundary Condition

Similarly, one can represent a **specified diffusive mass flux condition** as (referring to Figure 1.43)

$$q_{Dx}|_{i,n+1}. = \zeta D|_B \frac{c_{i+1,n+1} - c_{i,n+1}}{\Delta z_B} + (1-\zeta) D|_A \frac{c_{i,n+1} - c_{i-1,n+1}}{\Delta z_A} = q_{i0} \quad (2.54)$$

where ζ plays the role of β in equation 2.44 and weights the upstream and downstream components of the derivatives surrounding the node i.

2.8.3 Third-Type Boundary Condition

Although up to this point, development of boundary conditions for the transport equation has paralleled that of the flow equation, in the development of the third-type boundary condition there occurs a divergence from the development of the condition described for the flow equation. The total-flux condition for the one-dimensional case is

$$\tilde{q}_{zi} = q_{zi} c - D_{zz} \frac{\partial c}{\partial z} \quad (2.55)$$

which has no obvious counterpart in the flow equation. Careful examination reveals that this is a third-type or **Robbins** boundary condition. One would represent this condition as

$$\begin{aligned}\tilde{q}_{zi} &= q_{zi} c_{i,n+1} - \left[\zeta D|_B \frac{c_{i+1,n+1} - c_{i,n+1}}{\Delta z_B} + (1-\zeta) D|_A \frac{c_{i,n+1} - c_{i-1,n+1}}{\Delta z_A} \right] \\ &= \tilde{q}_{zi0} \end{aligned} \quad (2.56)$$

where ζ is the spatial-derivative weighting coefficient.

2.9 INITIAL CONDITIONS

Since the transport equation describes the transient behavior of contaminant concentration, initial concentration values are required. The initial conditions for transport are usually specified as zero concentration everywhere except at the contaminant source. At the source a first- or second-type boundary condition that specifies either the concentration or mass flux, respectively, is provided.

It is tempting to use hand- or machine-contoured concentration plumes as initial conditions to a model. Using such concentrations is not advised because generally these contoured plumes do not conserve mass since they were generated without any dependency on the physical equations governing the system. If these contoured plumes are used, the model will change the initial conditions in an effort to generate a mass-conservative plume. The analyst should be aware of this behavior in interpreting the simulated plume behavior.

2.10 MODEL PARAMETERS

The **dispersion coefficient D** in the transport equation plays a role similar to that of the hydraulic conductivity **K** in the groundwater flow equation; however there is one important difference. The functional form of the dispersion coefficient depends on the groundwater velocity as seen in equation 2.47 and reproduced here as

$$D_{\alpha\beta} = \alpha_T |q| \delta_{\alpha\beta} + (\alpha_L - \alpha_T) q_\alpha q_\beta / |q| + D_m \delta_{\alpha\beta} \tag{2.57}$$

Recall that in the case of the hydraulic conductivity, one can often select the coordinate system that will make the coordinate axes collinear with the principal directions of the hydraulic conductivity tensor, thus eliminating the off-diagonal terms. In the case of the dispersion coefficient, however, this is not the case. The fact that the velocity varies spatially means that the off-diagonal terms of the dispersion coefficient are always present. The consequence is that the dispersive flux vector has the form

$$\mathbf{q}_D = -\mathbf{D} \cdot \nabla c \tag{2.58}$$

which in expanded form yields

$$\begin{Bmatrix} q_{Dx} \\ q_{Dy} \\ q_{Dz} \end{Bmatrix} = - \begin{bmatrix} D_{xx} & D_{xy} & D_{xz} \\ D_{yx} & D_{yy} & D_{yz} \\ D_{zx} & D_{zy} & D_{zz} \end{bmatrix} \begin{Bmatrix} \dfrac{\partial c}{\partial x} \\ \dfrac{\partial c}{\partial y} \\ \dfrac{\partial c}{\partial z} \end{Bmatrix} \tag{2.59}$$

Let us now further expand and differentiate with respect to x one term of the vector on the left-hand side of equation 2.59:

$$\frac{\partial q_{Dx}}{\partial x} = -\left(D_{xx} \frac{\partial^2 c}{\partial x^2} + D_{xy} \frac{\partial^2 c}{\partial x \partial y} + D_{xz} \frac{\partial^2 c}{\partial x \partial z} \right) \tag{2.60}$$

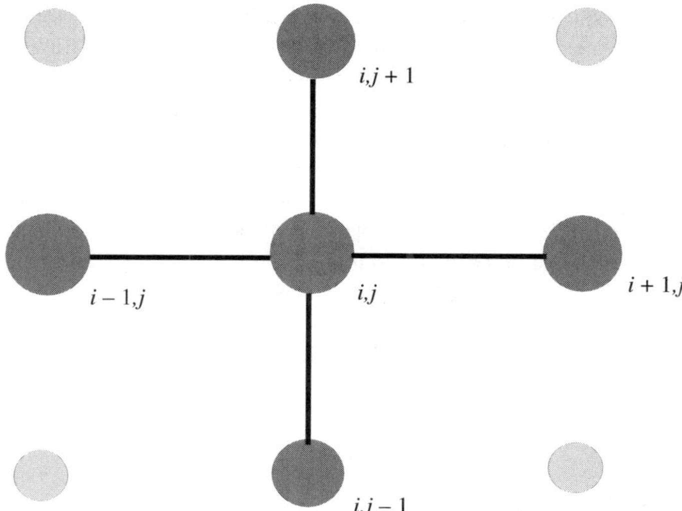

FIGURE 2.11. *Finite-difference template showing five-point (dark circles) and nine-point (all circles) nodal arrangements. Rectangular finite elements naturally generate the nine-point template.*

We see from this expression that it will, in general, be necessary to approximate a cross-derivative.

As they are normally applied, finite-difference methods are not suitable for the approximation of cross-derivatives. More specifically, finite-difference methods, in two dimensions for example, typically employ what is called a *five-point template* (Figure 2.11).

The five-point template consists of the five nodes that are connected by straight lines in Figure 2.11. These five nodes will represent, to order $(\Delta x)^2$, all the second-order derivatives except the cross-derivative. Representation of the cross-derivatives requires nine nodes, the normal five-point template plus the four corner nodes shown in Figure 2.11 that are not connected by straight lines. The nine-point template is seldom used in practice because of the resultant larger matrix bandwidth and consequent increase in computational effort required to solve the algebraic equations.

Rectangular finite elements naturally generate a nine-point template. Triangular finite elements generate an irregular template such that the number of nodes in the template depends on the finite-element arrangement. Thus, in general, unless a special finite-difference template is used, the cross-derivatives generated by the transport equation are more easily represented by finite-element equations than by finite-difference equations. When finite-difference methods are used to represent cross-derivatives, information at the corner nodes is often lagged by one time step. This normally makes the algorithm conditionally stable.

As in the case of hydraulic conductivity as presented in the flow equation, the dispersion coefficient in the transport equation must be defined at each node. The parameters that must be specified are the **longitudinal and transverse dispersivities**. The groundwater velocity, determined from the solution to the flow equation, is then used to compute the dispersion coefficient at each node.

The convective term in the transport equations requires that the groundwater flow velocity be specified over each element. The velocity is obtained via Darcy's law from the head (or water-table elevation) values calculated from the flow equation. Since triangular elements generate three head values, one at each node, they together define a plane over each element. The slope of a plane, in this case the head gradient, is a constant. Thus the velocity over a triangular element is a constant. In rectangular elements, the velocity is determined from four nodes and can vary linearly over the element.

In addition to the dispersivity and the velocity, the **porosity** and the **retardation coefficient** must be specified for the entire model. These parameters can be defined either by node or by element.

2.11 MODEL STRESSES

The *model stresses* associated with the transport equation are essentially those identified earlier for the flow equation. **Rainfall, pumping well discharge**, and **pumping well recharge** are the stresses normally encountered. Recall that in the transport model, the key term describing recharge is that expressed as $Q(c - c_0)$, where Q is the recharge or discharge rate and c_0 is the concentration of the fluid associated with Q. While one might be tempted to introduce the product Qc_0, which is the mass flux, into the model, this value, which has the units of mass per unit time, is seldom known in practice. It is more common to know the concentration of the fluid c_0 being recharged or discharged and the volumetric recharge or discharge Q. Given these two inputs, the model will determine and implement the mass flux input term.

Note that the term $Q(c - c_0)$ has an attractive mathematical property. If the concentration in the aquifer c reaches that of the recharging fluid c_0, this term vanishes. This imposes a natural constraint on the maximum concentration that can be generated in the aquifer. Interestingly, this term vanishes from the equation when there is fluid discharge, the reason for this being that the resident fluid c has the same concentration as the exiting fluid c_0. The discharging term appears to have no impact on the transport equation. The influence of the discharge still exists and comes from the fact that the discharging well affects the velocity field, which still appears in the transport equation in the guise of the convective term $\mathbf{q} \cdot \nabla c$.

2.12 RUNNING THE MODEL

Although the flow model is often run independently of the transport model, the obverse is not true. In transport simulation one normally runs the flow and transport model together so that the velocity field defined by the flow model evolves in time in parallel with the transport. An alternative approach that may be applicable when the flow field is not changing markedly over time is to save the various flow fields generated under different sequential physical circumstances and call on these flow fields from the transport model as they are needed. This will avoid recalculation of a velocity field derived from a previously adequately calibrated flow model.

The calculations performed in solving the transport model are very similar to those required to solve the flow model. There are, however, a few differences. One difference involves the specification of the spatial increment. In general, a steep concentration front requires a small space step to prevent undesired oscillatory behavior in the solution. More specifically, the Peclet criteria, mentioned earlier; that is

$$P_e = \frac{q\,\Delta x}{D} \leq 2 \tag{2.61}$$

should be adhered to whenever possible. The multidimensional counterpart to equation 2.61 is to take the largest of the possible values of P_e in the various space dimensions when determining the constraint to be satisfied (i.e., $P_e \leq 2$).

While, in concept, the *PTC* algorithm is unconditionally stable for linear problems (no water-table conditions), truncation errors associated with large time steps can compromise the concentration solution. As a rule of thumb, a time step that adheres to a Courant criterion value no bigger than 1 is a safe bet:

$$C_r = \frac{q\,\Delta t}{\Delta x} \leq 1 \tag{2.62}$$

Once again this is a guideline and may sometimes be violated with impunity. If the concentration values calculated appear inappropriate, however, a reduction in time-step size is a relatively painless test to perform.

Tucson Example

The new parameter, beyond those defined previously for the flow model, that is required to simulate contaminant transport is the **dispersivity**. The challenges associated with measuring this parameter are well known. Thus this parameter is normally identified via the **calibration process**. In the calibration process an initial guess of this parameter is introduced into the model and the resulting simulated concentrations are compared to the prototype. To the degree that the two values vary, the dispersivity parameter is adjusted until there is acceptable correspondence between the calculated and measured concentrations. The window used for introduction of the dispersivity is shown in Figure 2.12. Note that as in the case of hydraulic conductivity, the **dispersion coefficient**, as defined in equation 2.57, is a **tensor** requiring nine coefficients in the two-dimensional case. Unlike the hydraulic conductivity however, the dispersion coefficient cannot be considered a constant everywhere, as is evident through an examination of equation 2.57. The existence of a spatially varying velocity in the equation defining the dispersion coefficient dictates that the dispersion coefficient also varies areally. Moreover, the fact that the velocity, and concomitantly the dispersion coefficient, change spatially precludes a simple global-coordinate transformation to eliminate the off-diagonal dispersivity-tensor coefficients. Thus, the dis-

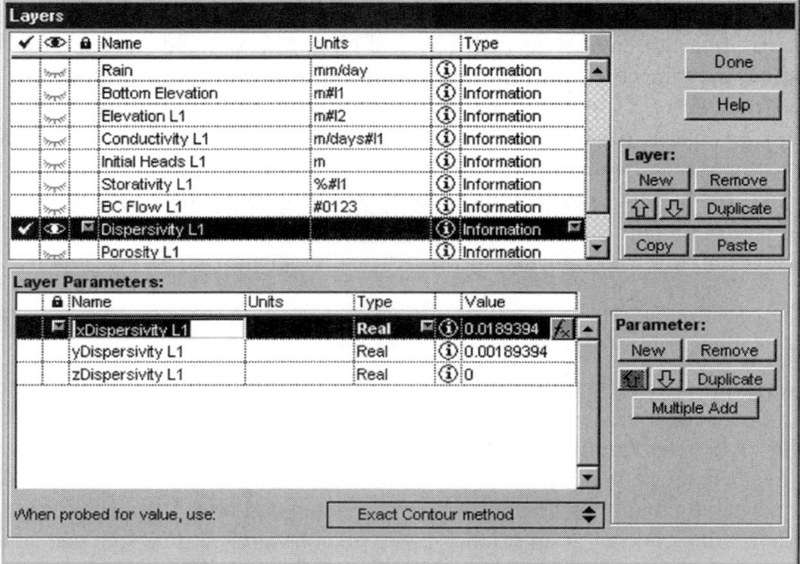

FIGURE 2.12. Layer window used to input the dispersivity values.

persion coefficient is calculated "on the fly" from the most current velocity values and the dispersivity values.

The window presented in Figure 2.12 is used to input the dispersivity in the three coordinate directions in a manner analogous to that used to introduce the **hydraulic conductivity** in Section 1.13. Having activated the *Dispersivity L1* layer, one selects the *Layers* option using the *View* menu. Information on dispersivity is introduced using the protocol used for hydraulic conductivity in Section 1.13.

2.13 OUTPUT

The output of concentration information is essentially the same as that for the hydraulic head discussed in Chapter 1. Contour maps of the concentration field are the most popular form of presentation. Because concentration values can range over several orders of magnitude, logarithmic rather than linear contours are often used.

Three-dimensional surfaces are an alternative method of representation of the concentration field. The resulting "fishnet" diagrams are particularly effective when combined with animation. Animation is especially effective for illustrating the past or future evolution of a contaminant plume.

Tucson Example

The **postprocessing** of concentration information is very similar to that used in Section 1.17 to generate hydraulic-head plots. Each step in importing the concentration

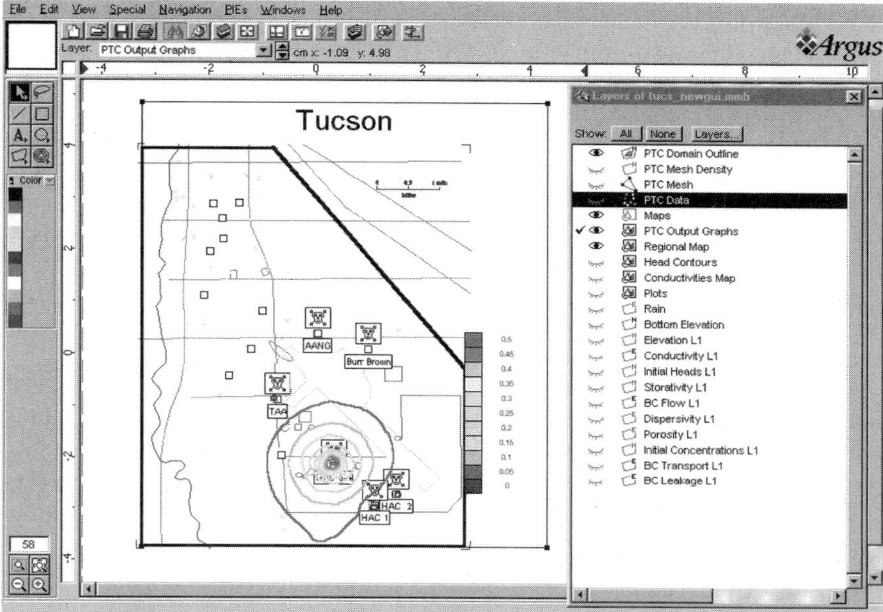

FIGURE 2.13. *Concentration plot generated from the transport simulation of the Tucson aquifer.*

data is analogous to the corresponding step in importing a hydraulic head. The name of the file to import for the last recorded transport calculation is now *Mesh_conc.fin*, in contrast to *Mesh_heads.fin* for hydraulic-head solutions. One may also wish to change the title associated with the plot to indicate that it represents a concentration solution. A concentration solution generated for the Tucson aquifer is presented Figure 2.13.

2.14 CALIBRATION

Much of the information provided in Section 1.18 is applicable to the calibration of transport. Because the accuracy of contaminant transport simulation depends very heavily on an accurate representation of the fluid velocity, and therefore the hydraulic-head distribution, head calibration is an important precursor to successful contaminant transport calibration. Of course, flow and contaminant calibration work hand in hand and should be considered simultaneously.

The calibration of a contaminant transport model involves comparing observed and computed groundwater concentration values obtained at wells or other borings. Data derived from spring discharge and surface-water bodies can also be helpful.

A comparison between observed and computed water-quality information is quite different from a similar comparison in the case of measured and computed head values. Because of the nature of the source concentration variability and the tortuous paths traversed by the contaminants in the subsurface, the concentrations observed at a well may be quite variable in time. On the other hand, the concentration values provided by a groundwater transport model are, by the nature of the averaging process inherent in the development of the governing partial-differential equations, averaged in time and space and therefore smooth. Thus the appropriate comparison is between the modeled-concentration results and a time- and space-averaged representation of the field concentration. For example, in Figure 2.14 a plot of contaminant concentration versus time is presented for a hypothetical series of observations. In the upper graph the point values of concentration over time vary significantly. In the lower graph a moving time average is used to smooth the data. In essence, the modeled-concentration values should be compared against the time-averaged concentration values. Similar logic is applicable to spatial averaging.

As in the case of groundwater-flow modeling, groundwater-transport model calibration involves varying field parameters, boundary conditions, and stresses within reasonable bounds in an effort to reproduce, via the model, the concentrations observed in the field. The two field parameters that will most influence the transport calibration are the hydraulic conductivity and the dispersivity. Porosity and retardation play a role similar to hydraulic conductivity inasmuch as the convective con-

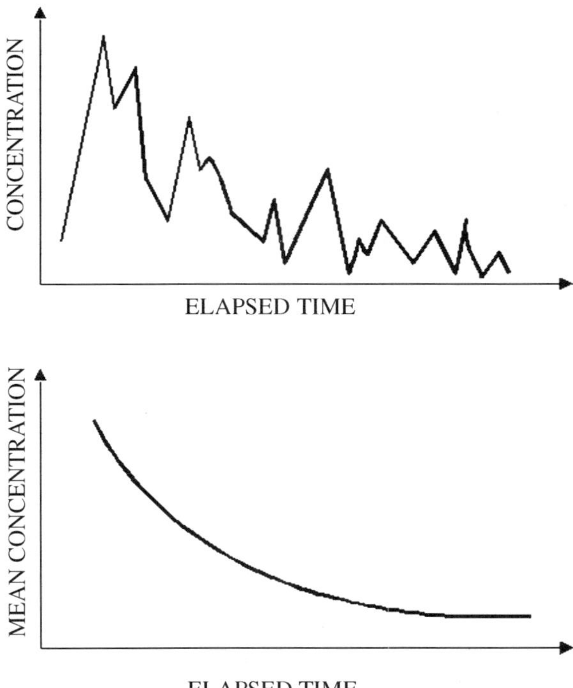

FIGURE 2.14. *An illustration of the concept of creating a moving time average concentration from fluctuating point measurement.*

taminant velocity is, in some sense, directly proportional to hydraulic conductivity and inversely proportional to porosity and retardation. While the porosity, hydraulic conductivity and retardation parameters may appear confounded in this instance, in practice this problem is ameliorated by the fact that the range of reasonable porosity and retardation values is relatively narrow compared to the range of possible hydraulic conductivity values. Thus the hydraulic conductivity tends to be the dominant factor of the three and most influences the success of the transport-calibration process.

Chemical or biochemical reactions, when present, may also affect the migration of contaminants. In fact, if a plume is observed to be largely immobile in a flowing groundwater system, the chemical reaction term can be an important factor in the system. When daughter degradation products are found along with the parent compound, the ratios of these compounds can provide a confirmation of the calibrated model. Due to the variability in the decay coefficients, it is difficult to use the ratios of daughter to parent compounds for initial calibration.

As noted earlier, dispersivity is very difficult to measure for a number of practical reasons. Therefore, this parameter tends to be a product of the transport calibration. In other words, the dispersivity value is determined by varying it within reasonable bounds until the spreading of the contaminant via dispersion in the model mimics that found in the field.

The overall level of contaminant concentration in a single contaminant-source problem is essentially proportional to the source concentration. In a model defined by first-type boundary conditions on concentration, if the source concentration is doubled, the concentration anywhere in the plume is also doubled. Once the overall concentration pattern is established, the source concentration can be used to adjust the overall magnitude of the concentration field.

2.15 PRODUCTION RUNS

The questions that can be addressed using the combined flow and transport model are much different from those considered with the flow model alone. With the addition of species transport, one can address a wide spectrum of groundwater-quality problems. Perhaps most important of these is the class of problems associated with groundwater contamination via human-made chemicals. The TCE-contamination case in Tucson, Arizona, that is outlined in the course of this book is a good example of this kind of problem.

Questions typically addressed with production runs using the flow and transport model involve either reconstructing past history or forecasting the future behavior of contaminant plumes. The reconstruction of contaminant plume history may be needed to determine contaminant source location. The approach is to assume the location of the contaminant source and then to simulate the behavior of the contaminant plume from the point in time when it was introduced into the groundwater until the point in time when the measurements defining the extent of the plume were taken. If a satisfactory match between the calculated and observed plume geometry is achieved, the hypothesized source location is a strong candidate for the sought-after source location. If there is not a satisfactory match, it is probable that either the location of the source is incorrect or the assumed time of introduction of the contaminant is in error. Should a match not be achieved, an additional run using an alternative source location or time of contaminant introduction must be made. This trial-and-error procedure is normally continued until a suitable source location and time of contaminant introduction are established.

Note that there is no general calibration of the contaminant transport aspects of the flow and transport model when it is used in this application. However, since the flow behavior in the model can be calibrated as described in Section 1.18, mass transport by convection can, in some sense, be calibrated even in this application.

Transport models can also be used to assess the effectiveness of plume containment and rehabilitation strategies. In this application a proposed containment or rehabilitation strategy is incorporated through specification of flow- and transport-model boundary conditions; More specifically, in the case of a pump-and-treat remediation strategy, the locations of the proposed remediation wells are represented in the flow model as second-type boundary conditions, that is, the well discharges are specified at the proposed well locations. Recharge-well water-flux rate can also be specified in the flow model and the recharge-water concentration specified in the transport model. The resulting model is run to simulate the behavior of the groundwater system over the remediation period and to evaluate the response of the groundwater system.

Based on the observed behavior of the water levels, the effectiveness of a containment strategy can be evaluated. In other words, the simulator will reveal whether the proposed inward-gradient constraints required to contain the plume have been satisfied.

If the remediation objectives are defined in terms of meeting specified concentration goals at locations specified at the end of the remediation period, the simulated concentration behavior under the influence of the proposed pump-and-treat system can assist in evaluating the efficacy of the proposed remediation strategy. The process normally involves locating the discharge and recharge wells and assigning pumping rates as identified in the proposed design. The behavior of the contaminant plume is simulated under these pumping conditions for the duration of the remediation period. Inspection of the simulated concentrations at the monitoring locations at the end of the simulation period reveals the efficacy of the design. It does not, however, reveal its efficiency. The efficiency is established by trying a variety of well locations and rates and examining the relative costs of those that satisfy the design constraints. Those found to be both effective and low cost are separated out for further consideration.

Design of both gradient- and concentration-based remediation alternatives can be automated. By combination of optimization algorithms and flow and transport models, least-cost remediation designs identified by the computer are possible.

Although we have focused above on pump-and-treat strategies, other remediation methods are amenable to examination using the same basic strategy. More specifically, one could include impermeable walls, drains, and impermeable caps in the suite of tools to be tested and evaluated by flow and transport modeling.

While groundwater contamination problems attributable to anthropogenic activities are common and deemed of considerable importance in the United States, saltwater intrusion into coastal aquifers is also of considerable importance worldwide. Such saltwater problems can be examined using the flow and transport codes. For that subset of saltwater intrusion problems that can be simulated using a concentration-independent density, the *PTC* code provided herein can be used. Using the simulator, potential problems can be identified and existing problems evaluated. In addition, the simulator can also be used to formulate and test the appropriateness and suitability of remediation design.

2.16 SUMMARY

The goal of this chapter was to extend the concepts on groundwater flow modeling introduced in Part 1 to include groundwater transport. The extension required introduction of the concepts of convection, dispersion, retardation, and chemical reactions. We demonstrated these modeling concepts via the Tucson-aquifer example. It was found that specification of boundary conditions, initial conditions, stresses, and aquifer parameters was similar, in concept, to the methodology introduced in Chapter 1 for flow simulation. In addition, we noted that the introduction of species transport into the model permitted the examination of a number of contaminant-related problems inaccessible using only flow modeling.

REFERENCES

[1] Huyakorn, P. S. and G. F. Pinder, *Computational Methods in Subsurface Flow*, Academic Press, San Diego, CA, 1983.

[2] Bedient, P. B., H. S. Rifai, and C. J. Newell, *Ground Water Contamination, Transport and Remediation*, Prentice Hall PTR, Upper Saddle River, NJ, 1999.

[3] Lorah, M. M., L. D. Olsen, B. L. Smith, M. A. Johnson, and W. B. Fleck, "Natural Attenuation of Chlorinated Volatile Organic Compounds in a Freshwater Tidal Wetland, Aberdeen Proving Ground, Maryland," *U.S. Geological Survey Water-Resources Investigations Report 97-4171*, 1997.

[4] Pinder, G. F. and W. G. Gray, *Finite Element Method in Surface and Subsurface Hydrology*, Academic Press, San Diego, CA, 1977.

Part 3

Finite-Element versus Finite-Difference Simulation

In this part we summarize the steps required to build a model. To demonstrate the similarity and differences associated with building a model using a finite-difference versus a finite-element based simulator, we illustrate each model-preparation step using both modeling strategies. To assure a consistent comparison, the Argus ONE interface is used to construct both models. *MODFLOW 96* is used in this analysis.We begin by modeling a straightforward hypothetical aquifer and then extend the analysis to the same real-world Tucson example considered earlier.

3.1 ELEMENTARY APPLICATION

3.1.1 Groundwater Flow

The first application will involve the simulation of groundwater flow. In a later section we tackle the more difficult problem of groundwater transport. The examples presented in this part were developed by Melissa McKay and Alexander Spiliotopoulos while they were graduate students in the College of Engineering at the University of Vermont. In the following development we present the finite-difference and finite-element approaches in parallel so as to provide the greatest insight into their respective strengths and weaknesses.

Upon startup of Argus ONE, the first window that appears is the one shown in Figure 3.1. Upon selecting the *PIEs* option, the indicated drop-down window or *combo box* is revealed. Among the many PIE options, the ones of interest to us are *New PTC Project* and *New MODFLOW Project*.

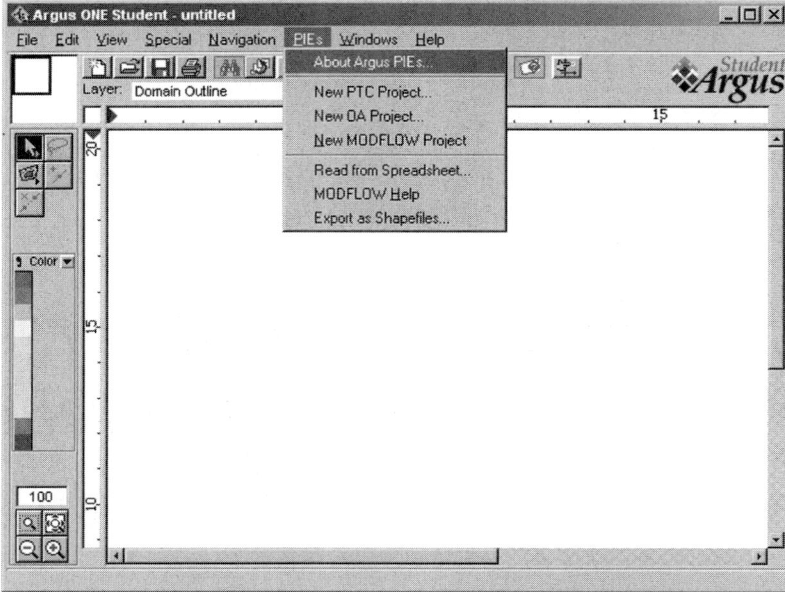

FIGURE 3.1. *First dialog window that appears on start-up of Argus ONE.*

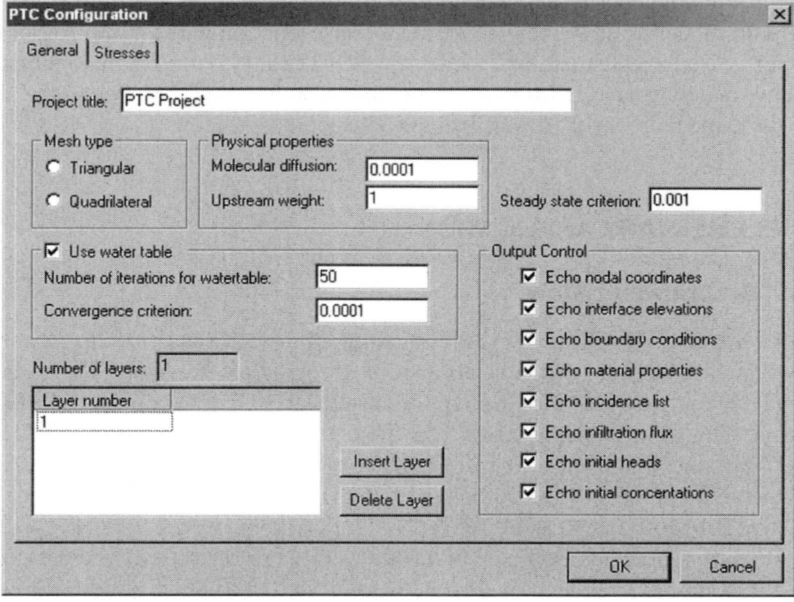

FIGURE 3.2. *Upon selection of the New PTC Project option one is provided with the PTC Configuration window.*

PTC Configuration Input We begin with the *PTC* option, which on selection reveals the dialog box presented in Figure 3.2. The *Configuration file* window, which is revealed by clicking on *PIEs* and then *New PTC Project...*, is shown in Figure 3.2. In the case illustrated in this figure, the *General* tab has been activated. The project name is provided by the user in the *Project title* text box. Designation of those input variables provided to *PTC* that the user wishes to have placed in files for later viewing is identified by activating the check boxes found in the *Output Control* frame. If a water-table problem is to be simulated, the box associated with the label *Use water table* must be checked and the maximum number of nonlinear iterations to be performed placed in the *Number of iterations for watertable* text box. The convergence criterion (in terms of water-level change) that must be met to accept the solution as accurate is placed in the *Steady state criterion* text box. The radio button associated with the type of finite element to be used must also be pushed in the *Mesh type* frame. The remaining text boxes are used when mass transport is to be considered and are addressed later. The list box records the number of layers in the proposed model. To add a layer, press the *Insert Layer* button, and to delete an existing layer press the *Delete Layer* button. In this example there is only one layer.

Activation of the *Stresses* tab in the *PTC Configuration* dialog box reveals (Figure 3.3) a list box and a series of text boxes that must be completed to further define the groundwater-flow problem to be considered. The list box is used to specify the number of stress periods and the various control attributes to be associated with each

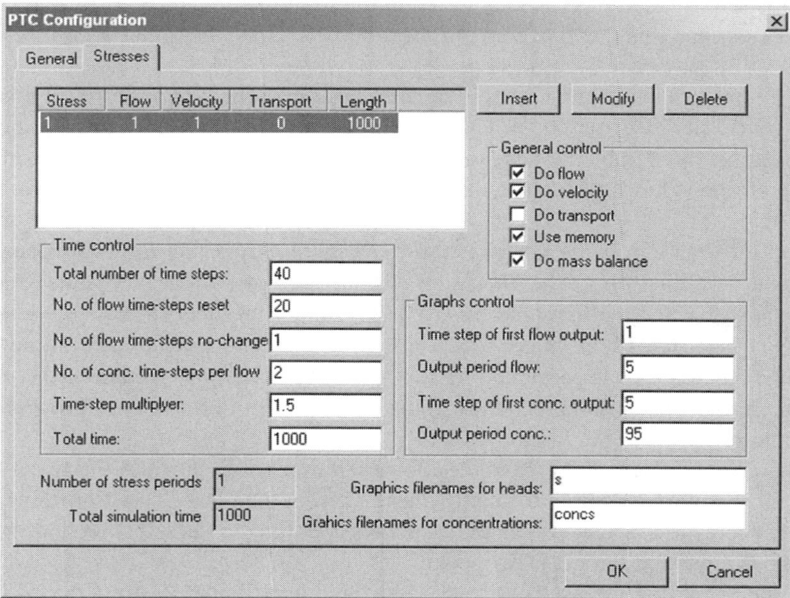

FIGURE 3.3. *Activation of the Stresses tab provides an opportunity to specify different General control attributes for each stress period to be considered (in this case one).*

stress period. In this circumstance, *Stress* is used to identify a period of time during which the various control parameters are to remain constant. For example, in this instance the numeral *1* appearing in the list box signifies that the corresponding control parameter is activated. More specifically, in this case the parameters *flow and velocity* have been activated, as can be confirmed by observation of the corresponding check boxes in the *General control* frame and the numeral *1* under each of these attribute headings in the large text box. *Transport* is not activated, as evidenced by the lack of a check mark in the corresponding check box and the numeral *0* under the *Transport* label.

The text boxes in the *Time control* frame are used to indicate the duration of the simulation, the number of time steps to be used, and any modification of the time-step size that may be desired. In the example shown in Figure 3.3, the total period of time to be simulated is *1000* time units, as indicated in the *Total time* text box. In specifying this number, the user has de facto selected the time unit to be used for the duration of the data input. If, for example, 1000 days are implied by the number 1000 in this text box, other input parameters that contain time units, such as hydraulic conductivity, must also be defined in terms of days.

The time step will be modified from one step to the next by the *Time-step multiplier*, in this case given by the number *1.5*. Use of this multiplier will cause each successive time step to be increased by 50 percent. If the user wishes to increase the time step less often, the number appearing in *No. of time-steps no-change* text box can be increased accordingly. When it is desired to reset the time step after a number of steps have been executed, the number of intervening time steps between resetting the time step to its initial value is placed in the *No. of flow time-steps reset* text box. In this example, this text box contains the value *20*. Thus, the time step is reset to its initial value after each 20 time steps.

The total number of time steps to be used in simulating the total time, in this case 1000 time units, is given by the value in the *Total number of time steps* text box. In this example the number of steps to be used is *40*. Since we are not now considering transport, the value appearing in the *No. of conc. time-steps per flow* text box is not important.

The information in the *Graphs control* frame controls the output of computed results to specified files. The *Time step of first flow output* text box value specifies the time step when the first output is to be saved. The number in the *Output period flow* text box states how many time steps must elapse between each output from the flow calculations. The remaining text boxes in this frame pertain to the output of computed concentrations and are considered later in our discussion of transport simulation.

The *Graphics filenames* text box allows the user to specify the file name to be used for the output of the information designated in the *Graphs control* frame. Due to idiosyncrasies associated with the Windows operating system, it is prudent to limit this file designation to one or two letters.

In the event that the user wishes to designate more than one stress period in the model, the *Insert* command button is used to create additional periods. The information requested in the list box is required for each new stress period. It is essential that the *Modify* command button be pushed for the changes in the list box to take effect! Pressing the *OK* command button does not automatically save the new list-box information.

MODFLOW Configuration Input The configuration windows associated with the input of information to the finite-difference program *MODFLOW* are presented in a series of figures beginning with Figure 3.4. Selection of the *Project* tab results in the presentation of four text boxes associated with the *Project Information for MODFLOW Simulation* frame. Information provided in the *Project name*, *Date*, *Description*, and *Project title* text boxes serve to identify the project and the associated output files. Note that in contrast to the *PTC Configuration* window, a *Help* control button is provided for interactive user help.

Activating the *Time* tab reveals the window shown in Figure 3.5. The information requested is analogous to that requested for *PTC* in the *Time control* frame of the *Stresses* window. Using the *Transient/Steady state flow* spin button, the *Transient flow(0)* option is selected. The *Time units* spin button allows for the selection of time units, in this case *Minutes(2)*. Finally, the number of stress periods to be considered is recorded in the *Number of stress periods* text box.

In the list box, information on the length of the simulation period [*Length (PERLEN)*], number of time steps [*No. of steps (NSTP)*], and the time-step multiplier [*Multiplier (TSMULT)*] are recorded. The actions taken here are equivalent to those taken earlier in development of the *PTC* data set. The major difference lies in

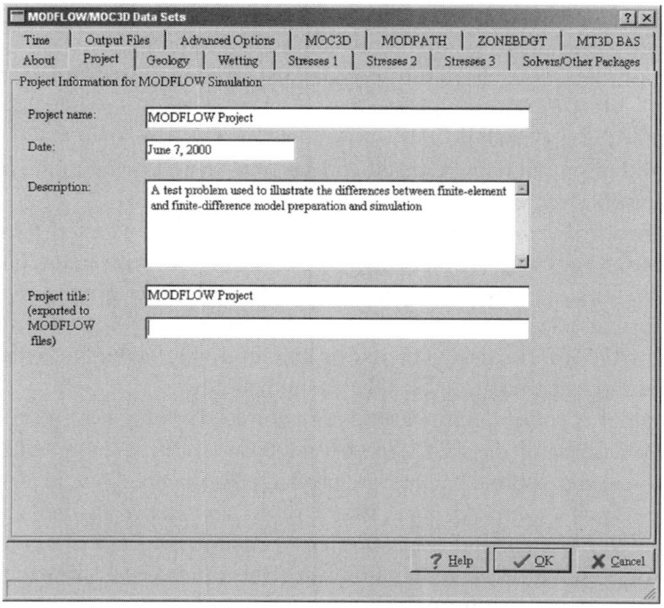

FIGURE 3.4. *Project information window used for MODFLOW configuration input.*

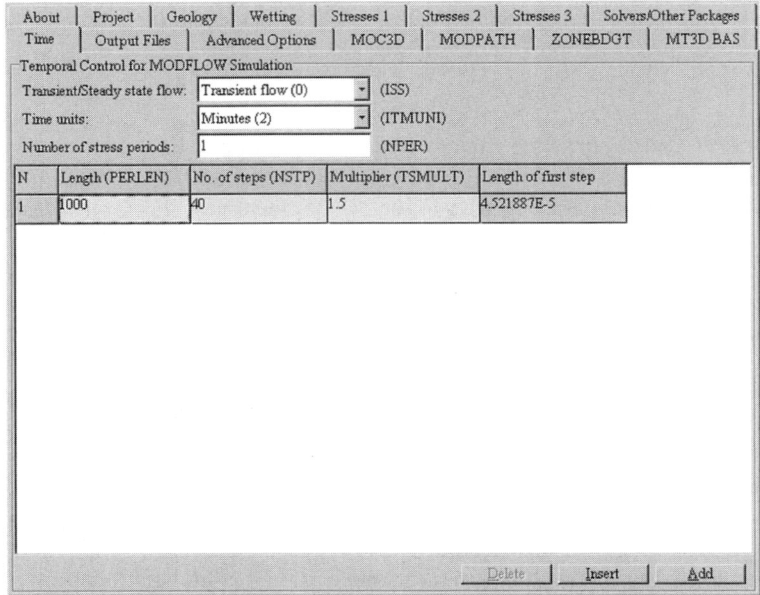

FIGURE 3.5. *Configuration window dedicated to the input of time-related information for MODFLOW simulation.*

the fact that included in the stress period information list box for *PTC* there is found information on the availability of velocity and transport options.

Selection of the *Output Files* tab activates the window found in Figure 3.6. The *Root name for MODFLOW Simulation files* text box is used to provide the file name to be associated with all *MODFLOW* output. The file name extension will be descriptive of the kind of output being created and saved to disk. For example, information pertaining to well discharge has the extension *.wel.*

The *External Files for Head/Drawdown Solution* frame is used to specify the format to be used for the files containing the head and drawdown data. For example, the *Export Head data* combo box permits selection of either *Formatted text file* or *Binary file* formats. The labels found in the *External Files for Cell-by-Cell Flows* pertain to how the information gathered on the fluid volume budget will be stored. Checking an associated box activates the desired strategy.

Printed output is controlled by a series of parameters similar to those used in the *Graphs control* frame of the *PTC Stresses* window. The frequency with which the head, drawdown, and volume balance are printed is determined by the *Head solution printed*, *Drawdown solution printed*, and *Overall volumetric budget* spin buttons. *None, Each N'th timestep*, and *Last timestep of each stress period* are the available options. Associated with each spin button is a series of text boxes in which the value of the parameter *N*, the line format, and the FORTRAN format control information are provided. In the *Output to external files* frame the same type of information is

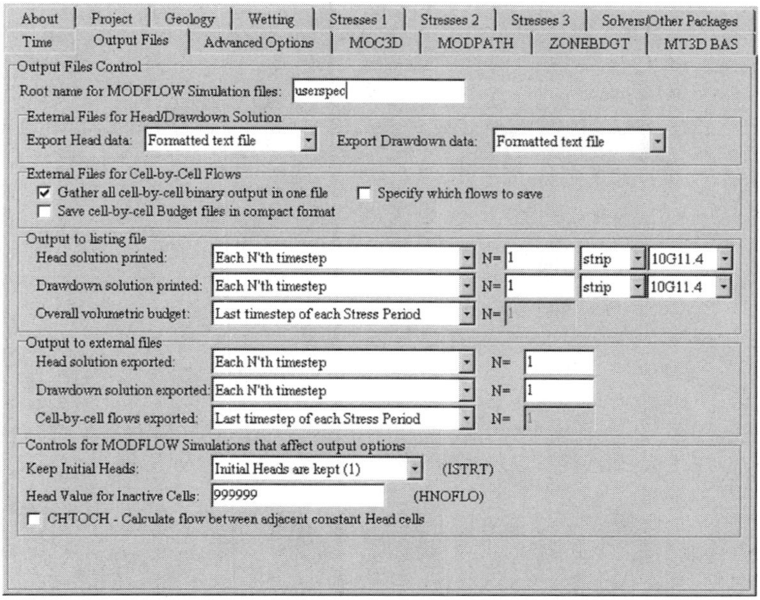

FIGURE 3.6. *The Output Files window is used to provide information regarding the files to be generated and their formats.*

provided for writing to files not used explicitly for printing. Although the counterpart for the *Output to listing file* frame in *MODFLOW* is found in part in the *Graphs control* frame of the *Stresses* tab window in *PTC*, there is no specific counterpart in *PTC* to the function defined in the *Output to external files* frame.

The *Controls for MODFLOW Simulations that affect output options* frame is used for miscellaneous control specifications. The *Keep Initial Heads* spin button allows the user to save the initial heads so that drawdown relative to the initial heads can be calculated. The *Head Value for Inactive Cells* text box is used to record a number that is used to define areas of the finite-difference mesh that are external to the active area. In other words, nodes where this number appears are considered external to the aquifer boundary. The *CHTOCH—Calculate flow between adjacent constant Head cells* check box is activated when flow between adjacent constant head cells is required, as is the case when simulating mass transport using *MOC3D*. There is no counterpart to this frame in the *PTC* GUI.

The *Geology* tab window appearing in Figure 3.7 is used to provide information on model layering. As in the case of the list box found on the *PTC General* tab window of the *PTC Configuration* window, new layer input is generated by pressing the *Insert* command button, and layers are removed by pressing the *Delete* button. In the case of the *Geology* window, each layer is provided with additional information. In the list box one can designate the name of the layer. In the example of Figure 3.7, the single layer represented is designated as the *Top Aquifer*. Under the label *Simulated*

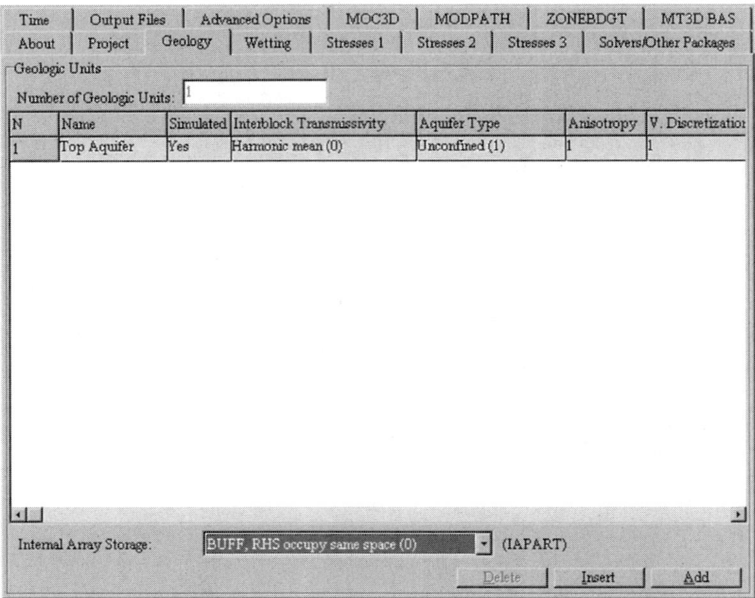

FIGURE 3.7. *Window used to input information on model layering.*

is recorded either a *Yes* or a *No* via a spin button that is revealed by double-clicking the underlying text box for each layer. A similar protocol is used to select the type of averaging to be used in computing the interblock transmissivities. Four types of averaging can be revealed via the spin box found by double-clicking the text box beneath the *Interblock Transmissivity* label.

Because *PTC* uses finite-element discretization in space, there is no counterpart to the averaging process that is used in finite-difference methods. Thus no averaging information is provided in the *PTC* input.

Determination as to whether a given layer is to be a confined, unconfined, or transitional aquifer is made in the configuration window option shown in Figure 3.7. Only the top layer can have the unconfined designation. The selection is made via the spin button exposed by double-clicking the text box beneath the *Aquifer Type* label. A similar designation is made in the *PTC Configuration* window in the *Use water table* frame. By selecting the *Use water table* check box, the user is identifying the top layer to be unconfined. Note, however, that *PTC* does not have the ability to transition underlying model layers from confined to unconfined should the water-level elevation drop below that of the bottom of the top layer.

The anisotropy ratio of the hydraulic conductivity in the areal plane is given by the value placed in the text box underlying the label *Anisotropy*. The number provided represents the ratio of the hydraulic conductivity in the y direction to the value in the x direction. A value of unity indicates an isotropic aquifer in the areal plane. In *PTC* the anisotropy is introduced in another part of the input stream, and no specification of this parameter is given in the *PTC Configuration* window.

The text box below the label *V. Discretization* indicates the number of finite-difference layers to be used to approximate the geological layer specified. The layers representing a given geological unit will be equally spaced. In *PTC* the layers are defined on the basis of mathematical accuracy and not on the basis of geological properties. In other words, a geological horizon could dip in such a way as to cross horizontally oriented model layers. Thus vertical discretization in *PTC* is accomplished simply by adding model layers via the *Layer number* frame of the *General* window of the *PTC Configuration* window (see Figure 3.2).

In Figure 3.8 we see continuation to the right of the information appearing in Figure 3.7. By double-clicking the text box below the label *Specify T*, the user reveals a spin button that will permit him or her to specify whether to specify the transmissivity of a unit directly or calculate it from other parameters. A similar strategy is followed for specifying the vertical conductance (*Specify Vcont*) and the storage coefficient (*Specify sfl*). The *Internal Array Storage* spin button allows the user to designate the form of internal storage to use during program execution.

Treatment of the dynamics of the water table is quite different in the *PTC* algorithm than that used in *MODFLOW*. In *PTC* a formal mathematical statement of the dynamics of the water table is used which requires no parameters other than the physical attributes of the system.

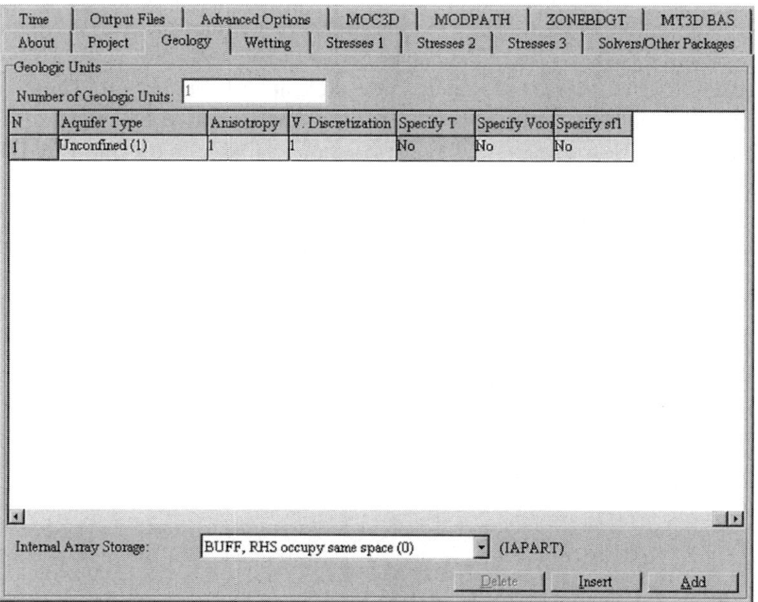

FIGURE 3.8. *Window used to input information on model layering (continued).*

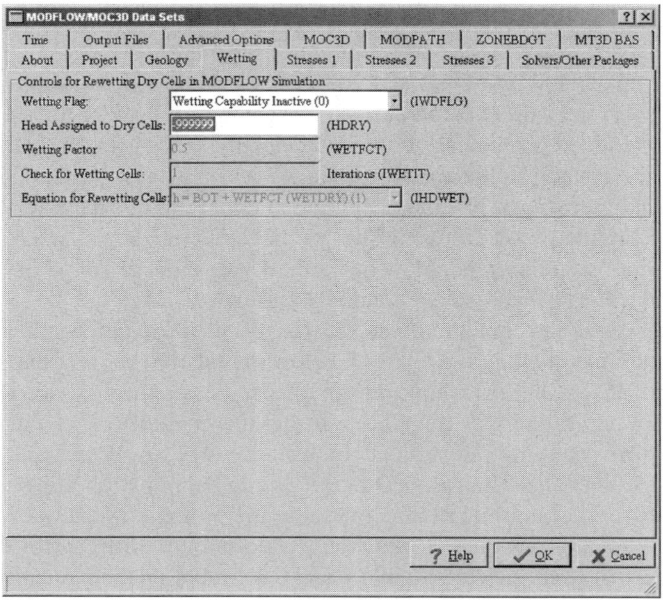

FIGURE 3.9. *Information utilized by MODFLOW to assist in representation of the the water table dynamics.*

MODFLOW, on the other hand, employs an alternative strategy for representing water-table dynamics that makes effective use of a number of heuristic parameters that can optimally be selected to enhance performance and convergence of the nonlinear solution. The relevant parameters are presented in Figure 3.9. Their roles are subtle and the reader is referred to the information provided by pushing the *Help* command button. Note especially the information available in the *Stability Problems in MODFLOW* option.

The *Wetting Flag* spin button is used to activate the option to wet finite-difference cells that, due to drainage, have become dry. In other words, if the fluid potential in the aquifer drops below the bottom of a finite-difference cell, the cell is no longer capable of transmitting fluid and is turned off. If the fluid potential increases above the bottom of the turned-off cell, then under certain circumstances, it can be turned back on.

The *Head Assigned to Dry Cells* text box allows the user to specify a head value that will be associated with a cell, should it become dry. The *Equation for Rewetting Cells* spin button provides two equations that can be used to define the head. Fig-

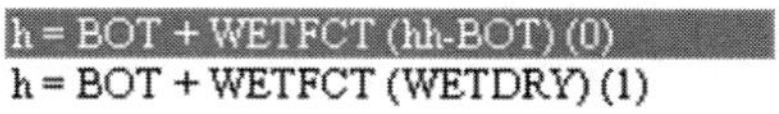

h = BOT + WETFCT (hh-BOT) (0)
h = BOT + WETFCT (WETDRY) (1)

FIGURE 3.10. *Equation for Rewetting Cells spin-button information on how the heads in rewetted cells will be determined.*

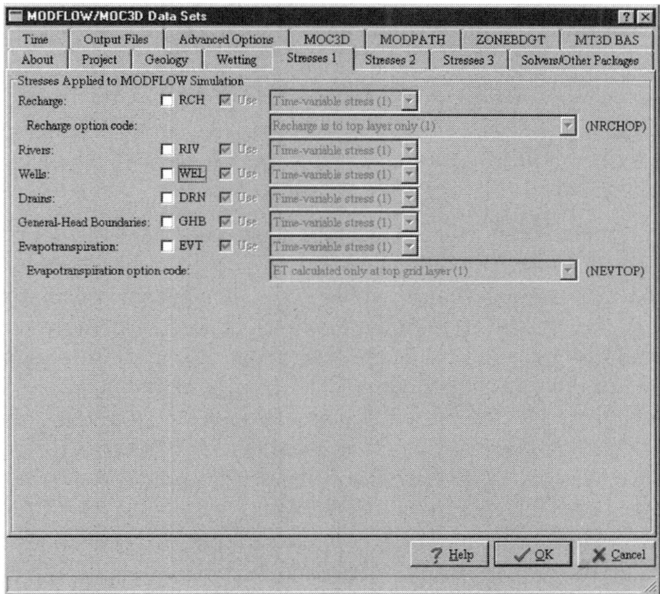

FIGURE 3.11. *Window provided by activation of the Stresses1 tab in the MODFLOW GUI.*

ure 3.10 presents these equations. Note that the wetting factor *WETFCT* appears in this equation along with other information provided as input elsewhere.

Consider now the *Stresses 1* tab of the window appearing in Figure 3.9. Selecting this tab reveals a series of hydrological-stress-related options (Figure 3.11). The various hydrologic stress options (packages) are activated by checking the corresponding boxes. Checking the *RCH* box activates two drop-down lists as viewed in Figure 3.12. As in the case for all the *Hydrologic Stress Options*, when the *Steady Stress (0)* option is selected, all stress information is assumed to be constant for the duration of the simulation period. When *Time-variable stress (1)* is selected, it is assumed that time-variable stress information will be provided for each stress period. The option *RCH* provides for the introduction of a specified recharge flux in units of volume per unit area per unit time.

Selection of the *Recharge option code:* spin button reveals the drop-down menu observed in Figure 3.13. The three options determine where in the model recharge is to be applied. In option one [*Recharge is to top layer only (1)*], recharge is to be applied uniformly only to the top layer of the model. Option *Vert. distribution in IRCH*

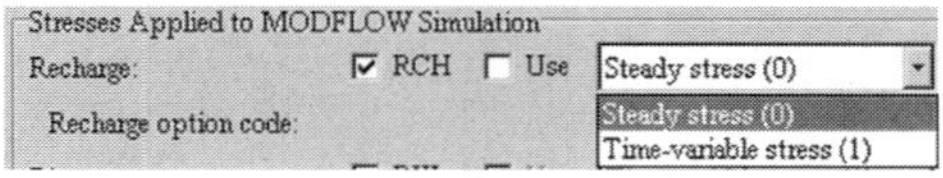

FIGURE 3.12. *Drop-down list activated by checking the RCH box.*

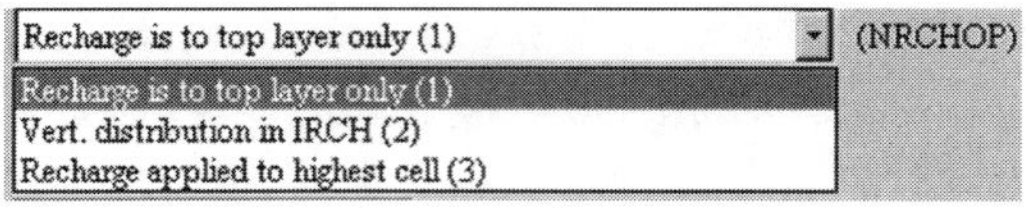

Recharge is to top layer only (1) (NRCHOP)
Recharge is to top layer only (1)
Vert. distribution in IRCH (2)
Recharge applied to highest cell (3)

FIGURE 3.13. *Drop-down menu presented on selection of the Recharge option code spin button.*

(2) allows for the specification of the recharge at any vertical location in any column. Specification of this information requires a spatially dependent information layer. In both of the options above recharge occurs only if the specified cell is active. The third option, *Recharge applied to highest cell (3)*, is used to circumvent difficulties that inactive cells might cause. As in the case of option one, no information layer is needed to specify this stress distribution. The *Use* check box adjacent to the *RCH* box can be unchecked to disable the functionality of the specified option without requiring that the data set prepared for the possible use of this option be removed.

Although we have discussed only one of the options provided by selection of the *Stresses 1* tab, there are five others (*Rivers*, *Wells*, *Drains*, *General-Head Boundaries*, *Evapotranspiration*). The various options are specific applications of classical boundary conditions. The *Recharge*, *Wells*, and *Evapotranspiration* options are, for example, second- or constant-flux-type boundary conditions. In the case of a well, a point boundary condition is applied.

The *General-Head Boundaries*, *Rivers*, and *Drains* are specific examples of third-type or Robbins boundary conditions. Sometimes this type of boundary condition is described as a *leakage condition*. It has the general form

$$au + b\frac{\partial u}{\partial n} = c \tag{3.1}$$

where u is the unknown function (e.g., hydraulic head); a, b, and c are constants; and n is the direction normal to the boundary of the region of interest. If we rewrite the equation as

$$\alpha(u - u_0) + \beta\frac{\partial u}{\partial n} = 0 \tag{3.2}$$

one can see that assuming that c is defined as the known head in a lake or river, the ratio β/α is assumed to be the resistance of flow from the river into the aquifer, and n is the direction normal to the stream bottom, equation 3.2 describes the flow into the aquifer due to the difference between the head in the aquifer u, and that in the river u_0. Another way to look at this equation is to recognize that the first term is, in some sense, a discrete form of Darcy's law, and the second is a continuous form. In any event, equation 3.2 can be used to represent the *General-Head Boundaries*, *Rivers*, and *Drains* options as defined by *MODFLOW*. It is worth noting that when β is zero, equation 3.2 describes a constant-head (first type) boundary condition and when α is zero we have a constant-flux (second type) condition.

We will see shortly that the corresponding input to *PTC* is far more basic. The various boundary conditions are specified simply as first-, second-, and third-type conditions. The user is required to determine which condition is appropriate for each hydrological situation. For example, the user is required to define a constant head boundary condition as a first-type condition and to specify the head value to be used by the model.

The *Stresses 2* tab provides access to the window presented in Figure 3.14. The purpose of this window is to provide information to the *Stream Package*. An augmentation of the input provided by the *River* and *Drains* packages, the *Stream Package* is a streamflow-routing package that allows for a more sophisticated representation of the physical behavior of open-channel flow than the *River Package*.

Also accessible from the window appearing in Figure 3.14 is the *Flow and Head Boundary Package*. By employing this package, it is possible to specify values of the head and inflow at various times within a stress period rather than just at the beginning. In essence this option allows one to modify the specified head and flux boundary conditions at each time step.

Activation of the *Stresses 3* tab reveals a window that provides for the specification of parametric input for the *Lake Package* and the *Seepage Package* (Figure 3.15). The *Lake Package* is provided to allow for dynamic changes in lake levels in response to changes in inflow and outflow. In essence the package performs a mass balance that considers runoff, rainfall, evapotranspiration stream flow, and groundwater recharge, from which it determines the lake stage. Lake stage is an important

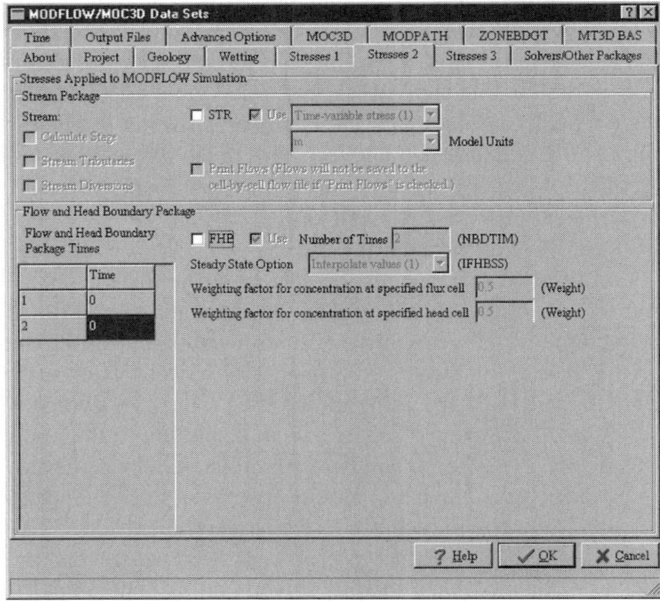

FIGURE 3.14. Window dedicated to the input of information on the MODFLOW Stream Package and Flow and Head Boundary Package.

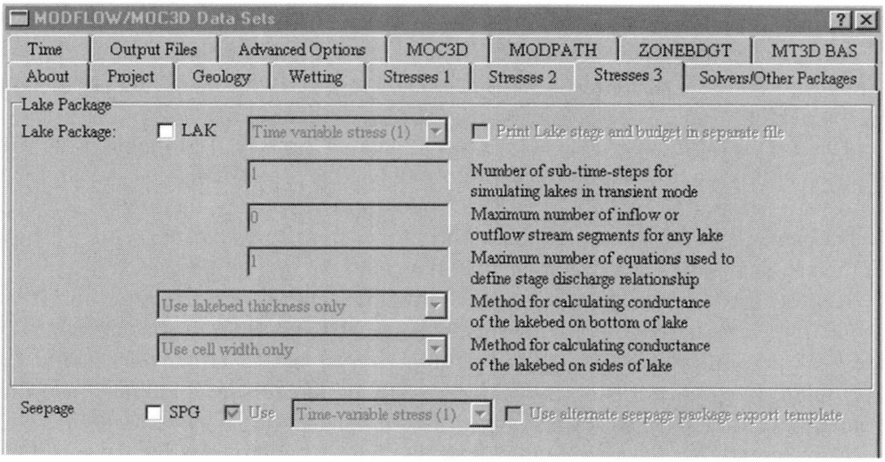

FIGURE 3.15. *Window dedicated to the introduction of parametric information on the Lake Package and the Seepage.*

element in the determination of aquifer recharge or discharge from surface-water bodies since it constitutes the driving force for flow into or out of an aquifer.

Selection of the *Solvers/Other Packages* tab reveals the window shown in Figure 3.16. At the top of this window is a check box to activate the *Horizontal Flow Barrier* package. The purpose of the package is to represent the occurrence of thin barriers of low hydraulic conductivity such as one would encounter in the case of low-hydraulic-conductivity magmatic intrusions or, in some instances, faults. The package modifies the conductance between grid cells whose boundaries coincide with these geological features so that very narrow elements that can lead to ill-conditioned coefficient matrices are avoided.

Below the *Flow Packages* frame in Figure 3.16 is the *Matrix Solution Method* dialog box. Four different matrix equation solvers are provided. The advantages to using one over the other depend primarily on the characteristics of the matrix generated by applying finite-difference approximations to the partial-differential equations describing, in this case, groundwater flow. In general, direct solvers such as *DE4* (the radio button highlighted in Figure 3.16) are the most robust. By **robust** we mean that is it is probable that when using a robust solver, a solution to the matrix equations generated from a properly posed groundwater problem will be achieved.

However, direct solvers, in general, require more computer memory to work efficiently than do iterative solvers such as *SIP* for large problems. *SIP* is also an option available in Figure 3.16. For very large problems, iterative solvers are often used. The subtitles that distinguish one iterative solver, and its concomitant convergence properties and accuracy, from another are beyond the scope of this book and the reader is referred to standard numerical texts for greater detail.

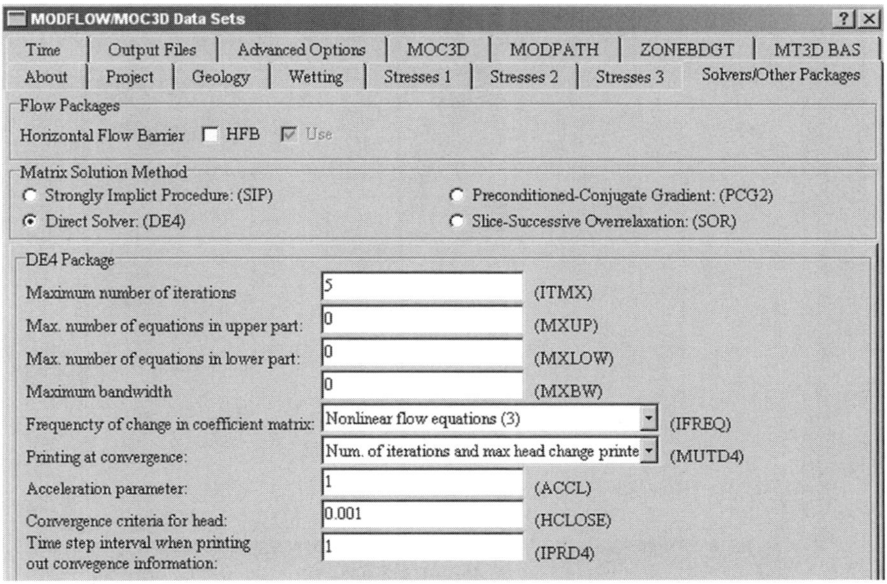

MODFLOW/MOC3D Data Sets

Time | Output Files | Advanced Options | MOC3D | MODPATH | ZONEBDGT | MT3D BAS
About | Project | Geology | Wetting | Stresses 1 | Stresses 2 | Stresses 3 | Solvers/Other Packages

Flow Packages
Horizontal Flow Barrier HFB Use

Matrix Solution Method
Strongly Implict Procedure: (SIP)
Preconditioned-Conjugate Gradient: (PCG2)
Direct Solver: (DE4)
Slice-Successive Overrelaxation: (SOR)

DE4 Package
Maximum number of iterations 5 (ITMX)
Max. number of equations in upper part: 0 (MXUP)
Max. number of equations in lower part: 0 (MXLOW)
Maximum bandwidth 0 (MXBW)
Frequencty of change in coefficient matrix: Nonlinear flow equations (3) (IFREQ)
Printing at convergence: Num. of iterations and max head change printe (MUTD4)
Acceleration parameter: 1 (ACCL)
Convergence criteria for head: 0.001 (HCLOSE)
Time step interval when printing out convegence information: 1 (IPRD4)

FIGURE 3.16. *Dialog box used to provide information for the Flow Packages and Matrix Solution Method text boxes.*

The issue of matrix-equation solvers does not arise in *PTC*. As described in Section 1.16, *PTC* uses a unique operator-splitting algorithm to approximate and solve the finite-element model efficiently. No other solution options are provided.

Spatial-Data Input

Model Geometry In this section we input the spatially dependent information that completes the input to the groundwater-flow model. To initiate this process, click on the layers icon found at the location of the cursor in Figure 3.17. A drop-down list box will appear, if it does not already exist, on the right-hand side of the dialog box. Several layers, in the sense of GIS layers, appear in this box. We must now address many of these layers, but we can disregard several because they pertain to groundwater transport, a topic we will consider shortly.

PTC As discussed briefly in Section 1.1, the first modeling step requires definition of the outer boundary of the model area. To activate the *PTC Domain Outline* layer, the area to the left of the "eye" in the *Layers of PTC* drop-down list box is clicked and a check mark appears. The geographic tool is now selected from the tool chest in the upper left-hand corner of the dialog box. The tool looks like a series of contours. Once clicked, the cursor is then moved to the workspace. The first click of the mouse locates the first vertice of the polygon that will be used to define the model boundary. One then clicks at each vertice until the polygon is complete. Termination of the

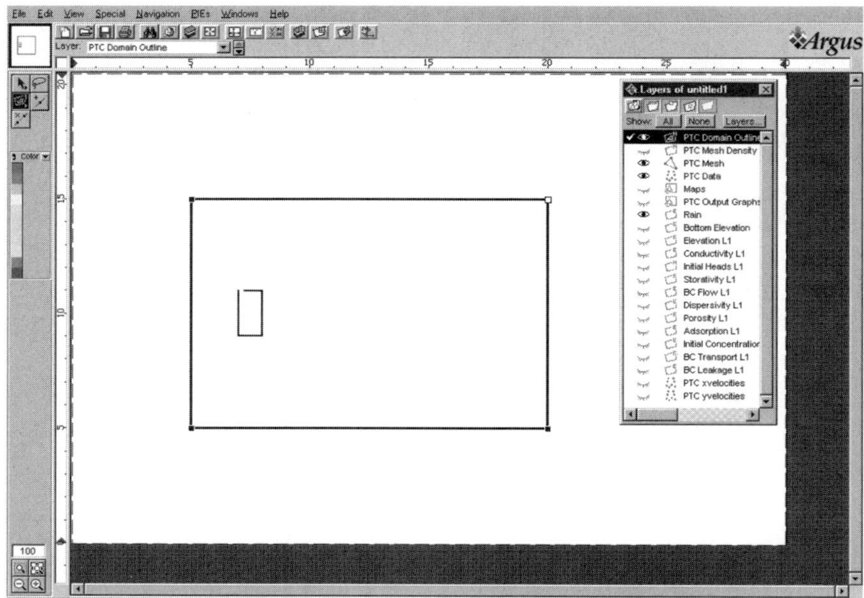

FIGURE 3.17. *Window used to define the physical domain of the finite-element model.*

drawing is realized by a double click, resulting in the appearance of the dialog box shown in Figure 3.17. In this problem we wish to have elements of approximately 0.4 unit. The length is established by considering the size of the model as determined by using the scales at the top and left-hand side of Figure 3.17. Of course, if triangular elements are used a variable mesh results and all triangles are not going to be of size 0.4 unit. *Density* refers to the average mean density or average length of elements to be used within the domain outline. In Figure 3.17 the outer square defines the perimeter of the model. Note that there is a small rectangle drawn inside the larger one. The purpose of the smaller rectangle is to define a region where a more dense finite-element mesh is to be constructed. Specification of this region at this point is a matter of convenience. The smaller region can be used to define a different mesh density, although at this stage we have selected the same 0.4 element length for this region (Figure 3.18).

MODFLOW To initiate the data input to the *MODFLOW* model, we first select the transport code we intend to use from the tabs appearing in Figure 3.4. In this example we use *MT3D* for transport. Thus we select the *MT3D BAS* tab and then *OK*. The dialog box shown in Figure 3.19 appears. The *MODFLOW Domain Outline* layer is activated, as was the case in the *PTC* input and the region of interest drawn using the same *Geographic* tool as with *PTC*.

Note that the *Layers* drop-down list box is similar to, but different from, that observed in Figure 3.17. This should come as no surprise, inasmuch as the information

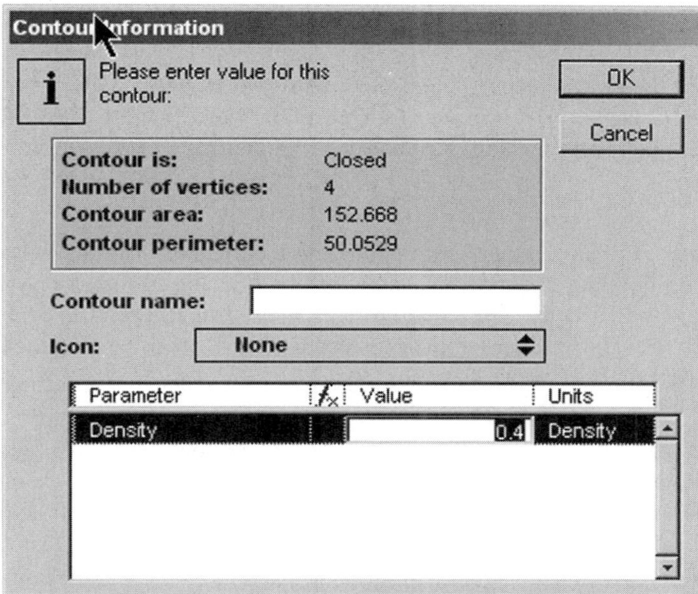

FIGURE 3.18. Dialog box used to specify mesh density for PTC.

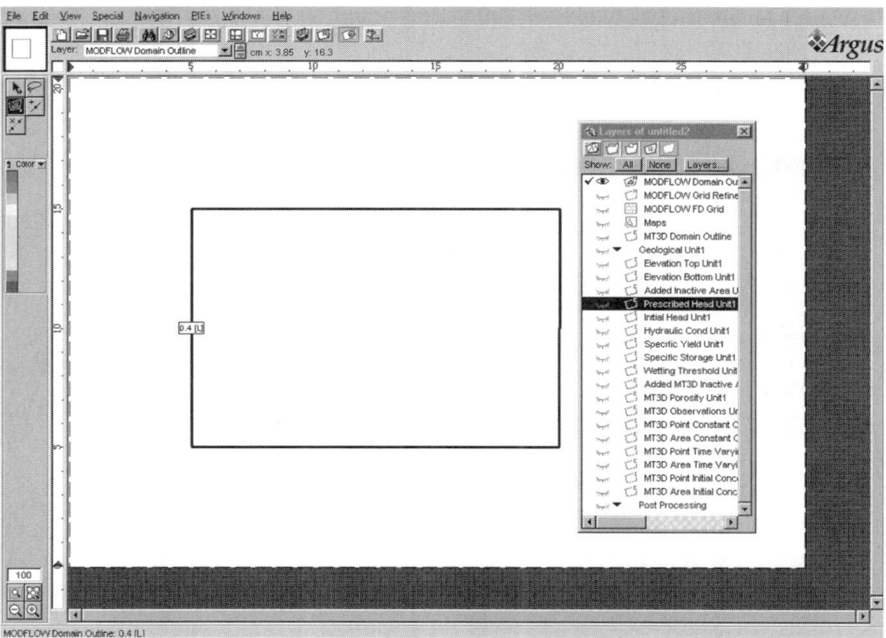

FIGURE 3.19. MODFLOW protocol used to define the domain outline for flow.

needed for the *MODFLOW* model is substantially more complex than that required by *PTC*.

The same protocol is used to define mesh density in *MODFLOW* as was described above for *PTC*. However, in the case of *MODFLOW*, which is finite-difference based, the mesh is much more regular and amenable to finite-difference cell-length specification.

Maps Layer In Part 1 a map was imported to facilitate the Tucson-model formulation. To do this, the *Maps* layer is activated by clicking to the left of the "eye" associated with that layer. The same protocol is used in *PTC* and in *MODFLOW*. However, in the very simple example considered here, no maps were needed or used. Thus the *Maps* layer is empty.

Formation Top and Bottom Elevations Several options are available to specify the top and bottom elevations. The simplest, and the one used in this example, is to use a global value for the entire model. To utilize this option, one first activates the *Bottom Elevation* layer (see the rightmost drop-down list in Figure 3.20). One then clicks the *Layers* button in the upper right-hand corner. From the drop-down list box, one selects *Layers*.... The dialog box on the left-hand side of Figure 3.20 appears.

To enter information on the global elevation, the f_x button is pushed (see the location of the cursor in Figure 3.20). This activates a dialog window that permits direct specification of numerical values, such as the zero used in this example, or other more sophisticated functions that are activated via the *Exact Contour method* spin button.

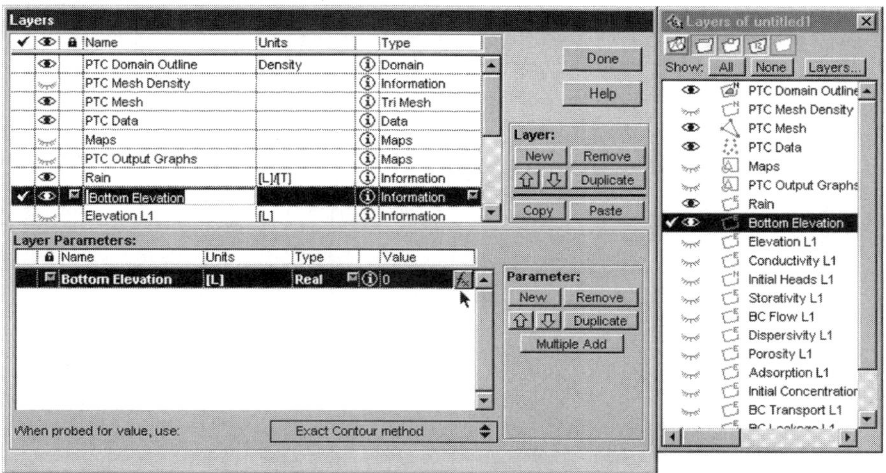

FIGURE 3.20. *Drop-down list and dialog box used to specify the elevations of geological formations.*

The top elevation of the top of the bottom-most layer is put into layer *Elevation L1*, which appears in the *Layers of PTC* drop-down list. Information for this layer is input using the same protocol as for the *Bottom Elevation*. The same protocol is used for specification of the formation elevations when using the *MODFLOW* GUI.

Note that if there had been more than one geological layer in the example, there would be a GIS layer for each geological horizon. For example, were there two layers, there would appear an *Elevation L2* GIS layer in the *Layers of PTC* drop-down list. The ordering of layers for *PTC* goes from the bottom (L1) to top (L2), whereas *MODFLOW* orders from top (L1) to bottom (L2).

Rainfall In most practical applications it is necessary to specify the rainfall, which is interpreted as a stress on the upper surface of the model. In this example, the rainfall is assigned a zero value and thus is input as a global variable. The protocol for accessing the dialog box for providing a global value of zero is shown in Figure 3.21. Note that the cursor indicates the text box where the value of zero rainfall appears. However, to enter this value, the f_x button is used to provide a dialog box suitable to input values in a number of different ways, similar to the case for formation elevations.

The units of input are length over time. Thus one may use inches per year or centimeters per year, depending on the units selected for the hydrological parameters, such as hydraulic conductivity. The same protocol is used to input rainfall to *MODFLOW* as to *PTC*.

Hydraulic Conductivity The same protocol is followed for the input of hydraulic conductivity as was used above to input formation elevations and rainfall. The appropriate dialog box and highlighted drop-down list are shown in Figure 3.22. In this

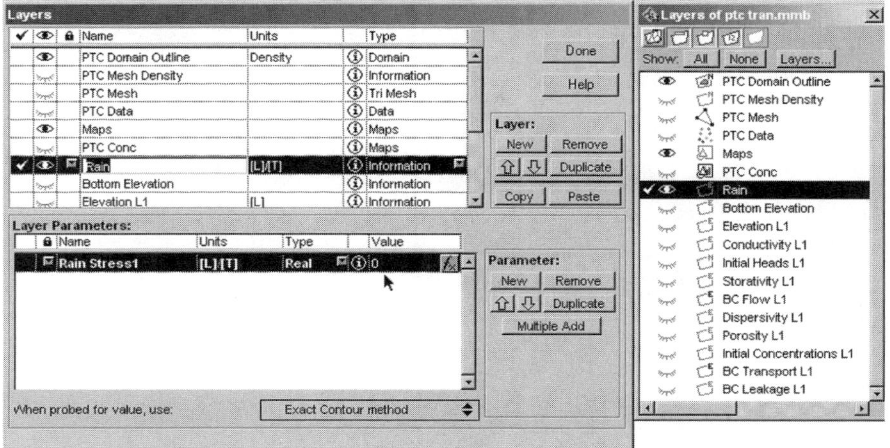

FIGURE 3.21. *Dialog box used to input a global value for rainfall. On the right is the Layers of ptc list box with the approriate Rain layer highlighted and activated.*

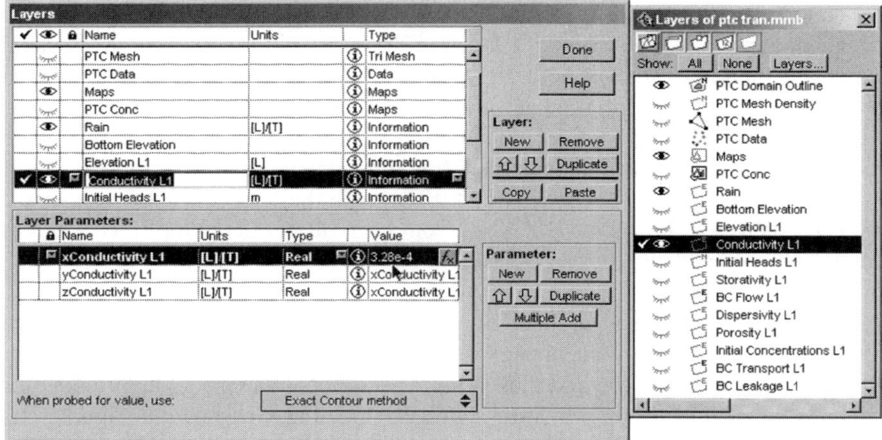

FIGURE 3.22. *Dialog box used to input values of hydraulic conductivity. To the right is located the drop-down list with the appropriate Layers of ptc layer highlighted and activated.*

case, however, there are three entries in the list box, one for each of the possible values of the hydraulic-conductivity tensor. In this example, only the conductivity value in the x-coordinate direction is specified. Once again it is assumed to be a constant over the entire model domain. The values in the y- and z-coordinate directions are indicated to be the same as in the x-coordinate direction. Had there been anisotropy in the model, the values in either the y- or z-coordinate direction would have been different and specified as a number.

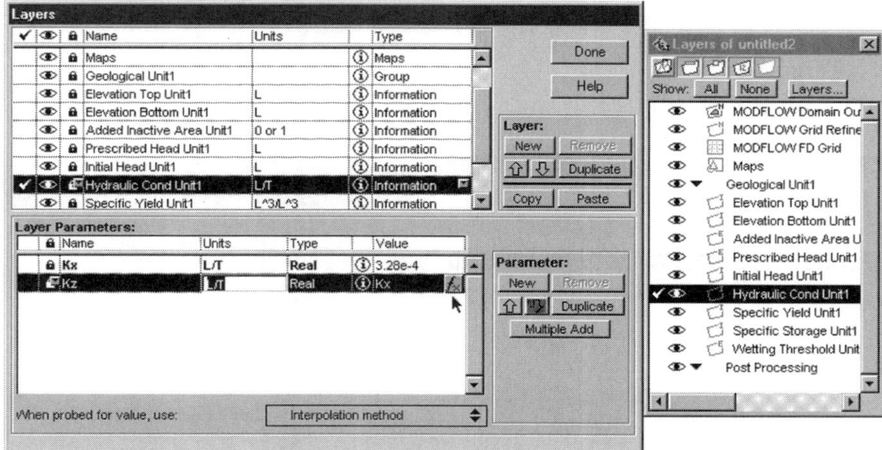

FIGURE 3.23. *Dialog box and drop-down list used to define the input of hydraulic conductivity and in MODFLOW.*

The protocol for *MODFLOW* is similar, but different from that presented above for *PTC*. As is evident in Figure 3.23, only the x- and z-coordinate values are specified. This is because the anisotropy between the x and y directions are specified in the window shown in Figure 3.7.

Specific Yield, Specific Storage, and Porosity The porosity and specific storage are introduced into the model in a manner analogous to hydraulic conductivity. However, whereas in the case of *MODFLOW* there are layers for each of these parameters, in the case of *PTC* there is only a *Storativity* (Specific Storage) layer. In the example problem considered here, the specific storage is 1.0×10^{-5} and the specific yield is 0.35.

Boundary Conditions Boundary conditions were discussed briefly earlier in this section and in Section 1.5.2. Recall that the boundary conditions are the points in the model along which we must define the behavior of the world external to the model. The practical realization of this requirement is a statement of the flow conditions along the boundary surfaces that define the model. We now show how one communicates these statements to the model via the Argus ONE interface.

In Figure 3.24 we present the information required to define a fixed-head boundary condition along the left-hand side of the model domain. The *Prescribed Head Unit1* layer is activated by checking it. The contour tool, which is located in the tool chest in the upper left-hand corner of this figure, is clicked. A rectangle is drawn that encompasses that segment of the boundary along which a known head is to be specified. Upon completion of the rectangle through a double-click on the last point location, the dialog box entitled *Contour Information* is activated. The value of the head specified is provided in the text box under the heading *Value*. The same procedure is used for each boundary segment on which a specified head boundary

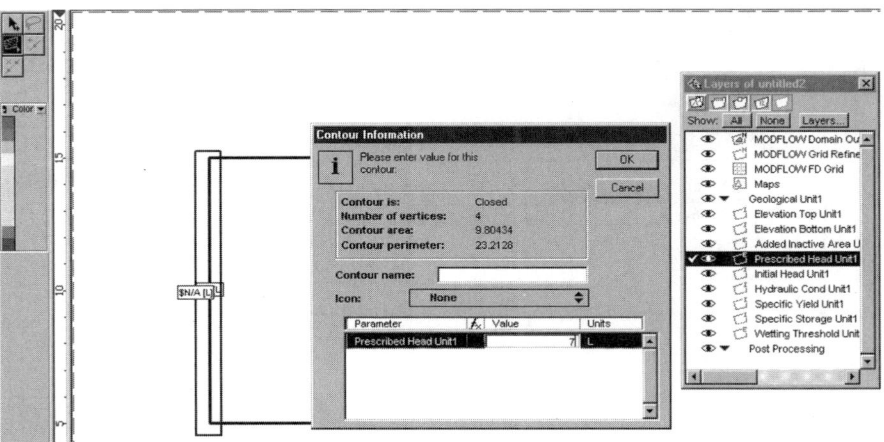

FIGURE 3.24. *Specification of constant-head boundary conditions in MODFLOW.*

condition is to be defined. Boundaries that are not addressed using this procedure are, by default, impermeable, no-flux boundaries. However, a specified flux can be defined using the boundary-condition specification protocol described earlier in this section.

The specification of boundary-condition information in *PTC* is achieved using a protocol very similar to that used by *MODFLOW*. Unlike *MODFLOW*, all three kinds of boundary conditions (i.e., specified head, specified flux, and leakage) are introduced at this point.

Consider the information provided in Figure 3.25 with the *BCFlow L1* layer checked. The segment of boundary along which a boundary condition is to be specified is defined using the *Contour Tool* in the same way as for *MODFLOW*. Upon completion of the defining rectangle, the dialog box entitled *Contour Information* in Figure 3.25 appears. Although similar to the corresponding dialog box generated by *MODFLOW* and shown in Figure 3.24, there is a subtle but important difference.

In the *PTC Contour Information* dialog box there are two text boxes to be addressed. The top one specifies the type of boundary condition and is entitled *BC TypeL1*. In this box one specifies either a 1 or a 2, which correspond to a type one (specified head) and type two (specified flux) boundary condition, respectively. In the text box identified with the label *BC Stress1*, the magnitude of the head or flux is given. Since we wish to define a specified head value of 7 along this boundary, as was the case for *MODFLOW*, the number 7 appears in this dialog box.

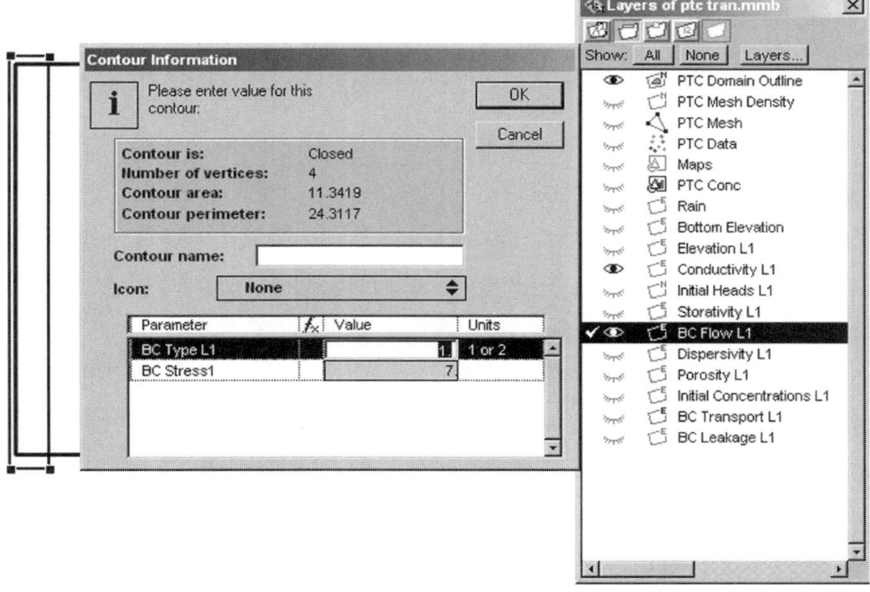

FIGURE 3.25. *Definition of boundary conditions using PTC. The box on the left defines the boundary segment of interest and is that to which the value selected (in this case, 7) is applied.*

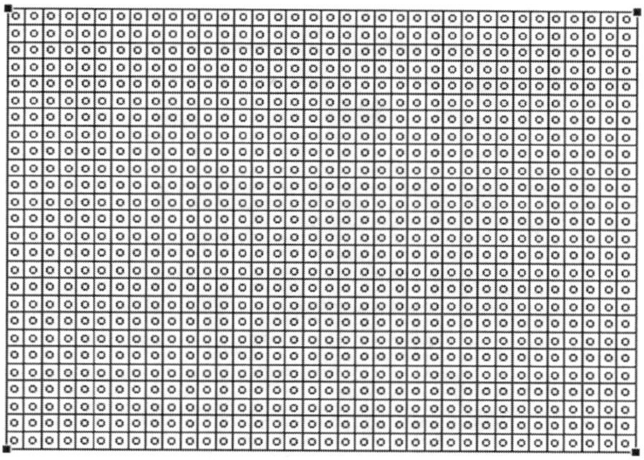

FIGURE 3.26. *Automatically generated finite-difference grid.*

To define a third type (leakage) boundary condition, the *BC Leakage L1* layer is activated and the protocol specified above for identifying the boundary segment of interest is employed. In this case a different dialog box is activated and the information regarding the required leakage conditions is specified. This topic is covered in greater detail and in Section 1.11.

Mesh Generation and Model Execution Having defined the necessary input data to both *MODFLOW* and *PTC*, one can now transfer the information to the respective modeling programs and simulate the system. To initiate *MODFLOW* data transfer and program execution, one first activates the *MODFLOW FD Grid* layer. The magic wand is selected from the toolbox in the upper left-hand corner. One must then click the mouse anywhere in the mesh and a finite-difference grid is automatically generated that is consistent with the *density* specification made earlier under "Spatial-Data Input: PTC" (see Figure 3.26).

From the main window, select *PIEs* and then select *Run MODFLOW/MOC3D* (see Figure 3.27). The resulting dialog box is shown in Figure 3.28. Note that for this example problem we have activated *MODFLOW 96* and the *Create input files and run MODFLOW* buttons. Output information for this simulation will be placed in the file *userspec.**, where the asterisk represents various file types to be used for postprocessing and error evaluation.

Selection of the *Model Paths* tab activates the dialog box presented in Figure 3.29. The purpose of this window is to allow the user to specify the location of the model executable. In this case the executable code is *MODFLW~1.EXE*, which is the DOS-compatible file originally named *Modflw96_Argus32.exe*. This is a very nice feature that *PTC* does not have. In the case of *PTC*, Argus ONE will look first in the local directory (the one from which *PTC* is called) and then through the various paths that you have specified in the *Environment Variables* on your PC. For a successful launch the *PTC* executable must appear in one of these locations. Upon clicking on *OK*,

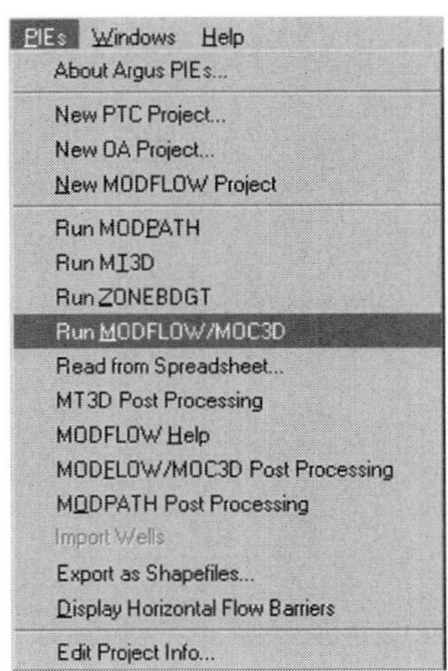

FIGURE 3.27. *Drop-down list used to initiate execution of MODFLOW.*

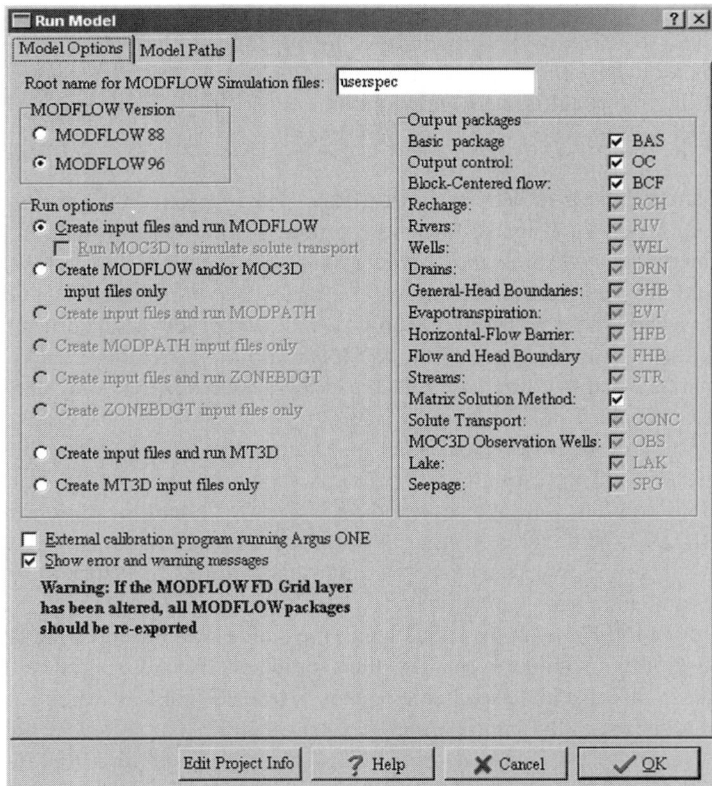

FIGURE 3.28. *Dialog box used to specify the nature of the simulation to be performed, in this case flow.*

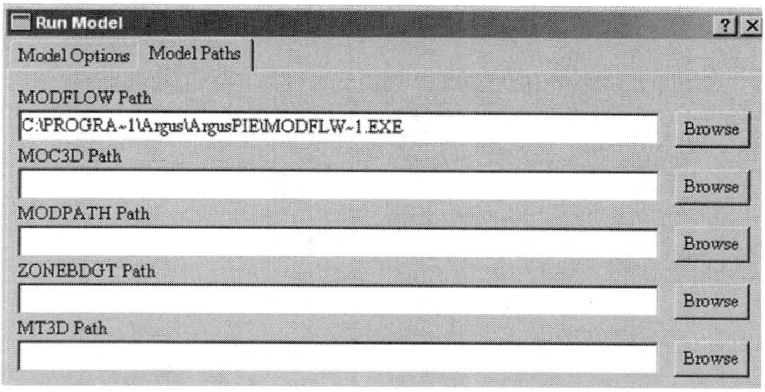

Run Model
Model Options | Model Paths
MODFLOW Path
C:\PROGRA~1\Argus\ArgusPIE\MODFLW~1.EXE
MOC3D Path
MODPATH Path
ZONEBDGT Path
MT3D Path
Browse

FIGURE 3.29. *Path specifications for the various executable files needed to run MODFLOW and its auxiliary programs.*

execution of the program begins. A window opens to inform the user that the program is *Processing Export Template*. A dialog window opens automatically to allow the user to define an export file name in the text box associated with *Enter Export File Name*. Upon closing this window, a window opens announcing *Export in Progress*. Finally, a DOS window informs the user that the model is running. Upon completion of a successful simulation, the DOS window appears as shown in Figure 3.30. Note that there is no message indicating successful completion of the simulation. When appropriate, a window containing a warning that irregularities have occurred appears and identifies the appropriate **.err* file in which the specific nature of the warnings or errors are recorded. The error file can be opened with a word processor such as Microsoft Word.

Flow simulation using *PTC* utilizes a protocol similar to that employed in launching *MODFLOW*. Upon activation of the *PTC Mesh* layer and clicking within the model domain with the magic wand, a finite-element grid consistent with the *density* specification is generated. The resulting mesh is shown in Figure 3.31. Note that there is slight distortion of the mesh near the rectangle located to the center left of the model domain. The reason for this distortion is that although not yet activated, the rectangle will define the concentration boundary conditions for contaminant transport, as described in the next section. The occurrence of such a boundary condition triggers an automatic modification in an otherwise uniform mesh.

Having activated the *PTC Mesh* layer and having selected the *PIEs* and *Run PTC* options from the main window (see Figure 3.32), one is presented with the dialog

```
C:\WINDOWS\System32\cmd.exe
C:\Program Files\Argus\ArgusPIE>C:\PROGRA~1\Argus\ArgusPIE\MODFLW~1.EXE /wait

C:\Program Files\Argus\ArgusPIE>Pause
Press any key to continue . . .
```

FIGURE 3.30. *Appearance of DOS window on the completion of a successful simulation.*

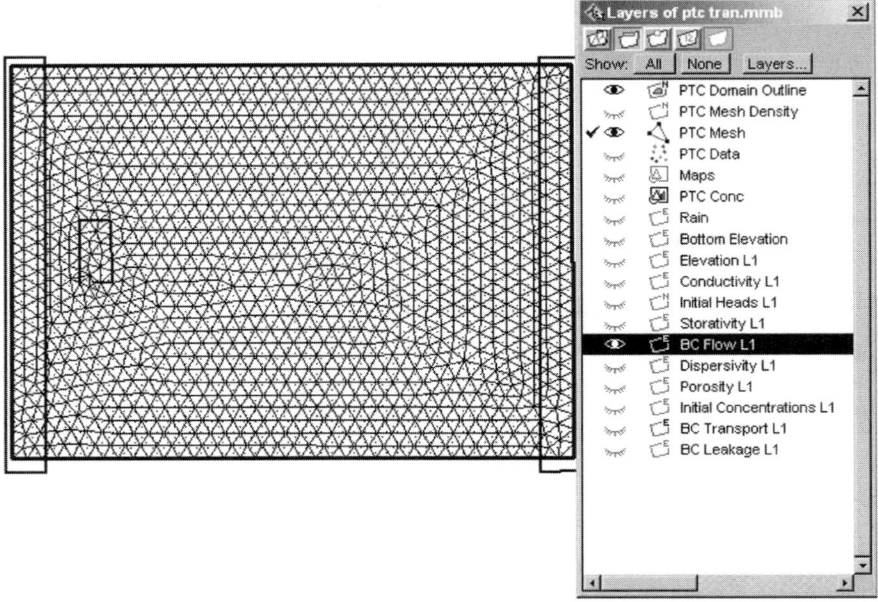

FIGURE 3.31. *Finite-element mesh generated prior to launching the PTC simulator for flow.*

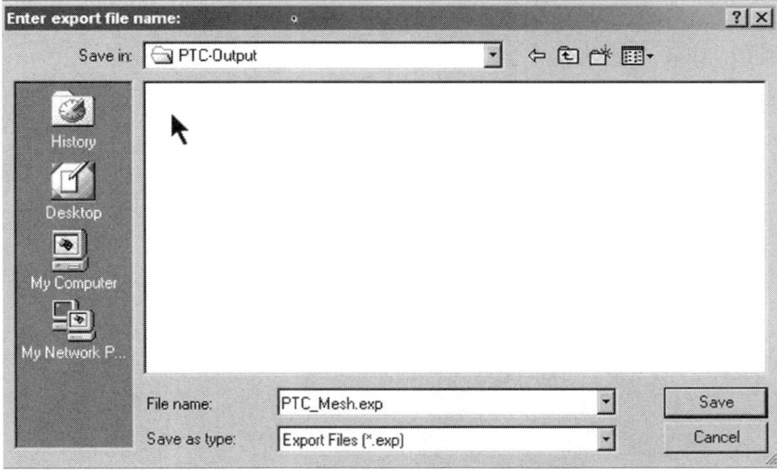

FIGURE 3.32. *Dialog box used to specify the file folder to which PTC will direct its output.*

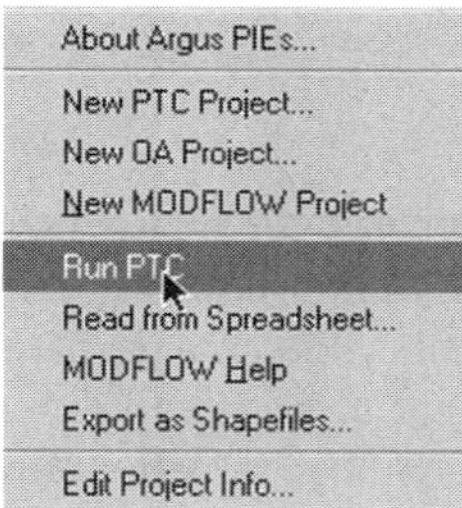

FIGURE 3.33. *Drop-down window activated by selection of the PIEs option with the PTC mesh layer active.*

```
C:\WINDOWS\System32\cmd.exe
C:\sw30\2000\MODBOOK-2000\MelissasStuff\TRANSPORTmmbs>ptc  0<PTC_Mesh.run
 completed flow time step number        1
 Open unit=      9 file=s_s1.1

 completed flow time step number        2

 completed flow time step number        3
 steady state flow reached at flow time step number        3

 completed flow time step number        4

 completed flow time step number        5

 completed flow time step number        6

 completed flow time step number        7

 completed flow time step number        8

 completed flow time step number        9

 completed flow time step number       10
C:\sw30\2000\MODBOOK-2000\MelissasStuff\TRANSPORTmmbs>pause
Press any key to continue . . .
```

FIGURE 3.34. *DOS window indicating progress of PTC simulation.*

window shown in Figure 3.33. From the drop-down list, select the folder to which *PTC* is to direct its output and click on *Save*. In Section 3.1.1, we selected the default folder. Here we specify that *PTC-Output* should be used. A window indicating that the export is in progress now appears.

Upon completing the export, *PTC* is launched. The resulting DOS window is shown in Figure 3.34. In contrast to the corresponding window in *MODFLOW*, the *PTC* DOS window provides a progress report on the *PTC* simulation.[1]. Examination of this window reveals that the completion of each time step is recorded. In addition, the point in time when the change in head value between successive time steps is within the tolerance used to define a steady-state system is provided. In this case steady state was achieved after the third time step. A pause action is used at the end

[1] Note that the output shown in the DOS window for PTC does not correspond to that obtained during the simulation. It has been truncated to show the entire simulation sequence

of the simulation, as was the case for *MODFLOW*, and hitting any key will take the user out of the DOS environment.

Postprocessing of Flow Simulation Results At this point we have obtained solutions for the model problem using both *MODFLOW* and *PTC*. Each program has produced a series of output files that contain the results of the simulation. *MODFLOW* and *PTC* are both capable of producing plots of the results; however, they utilize quite different protocols. We consider both in the following.

MODFLOW Postprocessing To initiate postprocessing of *MODFLOW* generated flow-simulation results, one selects the *PIEs* tab and then, from the resulting drop-down list, the *MODFLOW/MOC3D Post Processing* option, illustrated in Figure 3.35. The window presented in Figure 3.36 results. Since, at this stage, we have calculated only the groundwater elevations and not transport, the radio button identified as *MODFLOW (Head and Drawdown)* should be activated. Click on *Select Data Set* to identify the simulation results to be plotted. The list box shown in Figure 3.37 results.

Upon selection of the file presented in the list box in Figure 3.37, one obtains the information provided in Figure 3.38. The selection of output file and chart type using

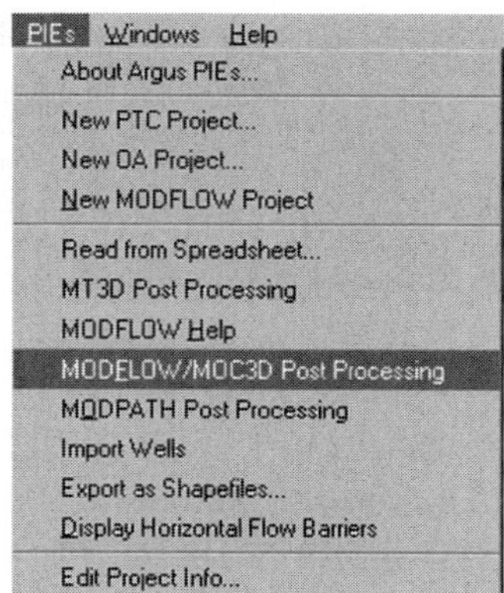

FIGURE 3.35. *Selection of post-processing option for MODFLOW.*

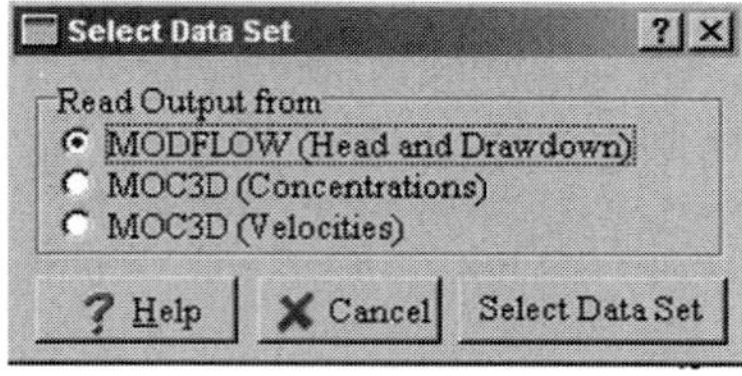

FIGURE 3.36. *Dialog box used to select the data set for the plotting of the MODFLOW head values.*

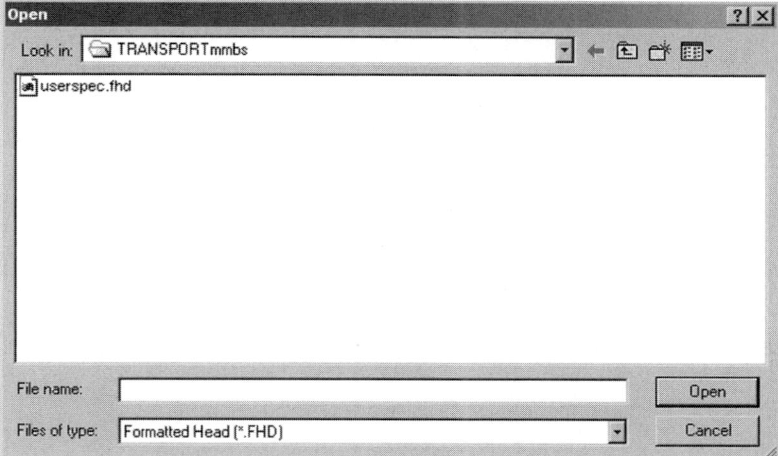

FIGURE 3.37. *List box used to select MODFLOW files to be post-processed.*

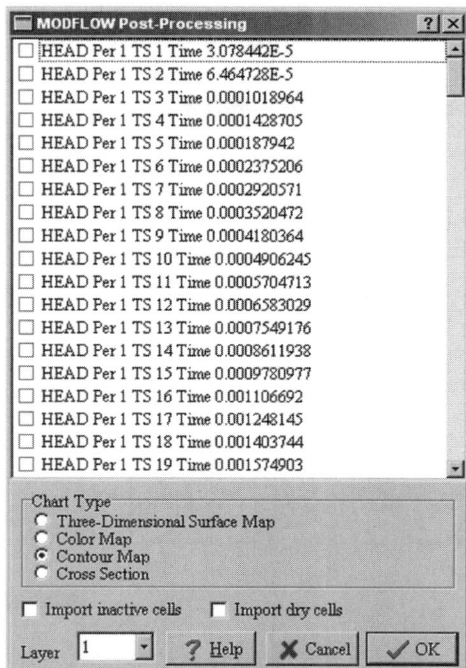

FIGURE 3.38. *List box provided to select specific MODFLOW files to be post-processed.*

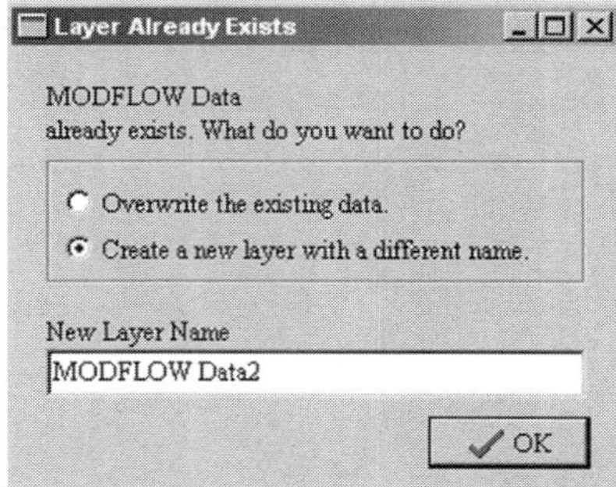

FIGURE 3.39. *Dialog box used to determine whether to overwrite an existing chart file or to create a new layer.*

the list box of Figure 3.38 is relatively straightforward. Simply check the box associated with the output period, time step, and elapsed time of interest. Also activate the radio button for the chart type of interest. In our example we wish to consider a contour map of the computed heads. From the resulting window shown in Figure 3.39, select the radio button that either overwrites an existing data set or creates a new layer. In our example we elect to create a new layer and name it *MODFLOW Data2*.

The resulting contour plot is shown in Figure 3.40. Note the layers that have been activated to produce this plot. In Figure 3.41 we see a magnified view of the head

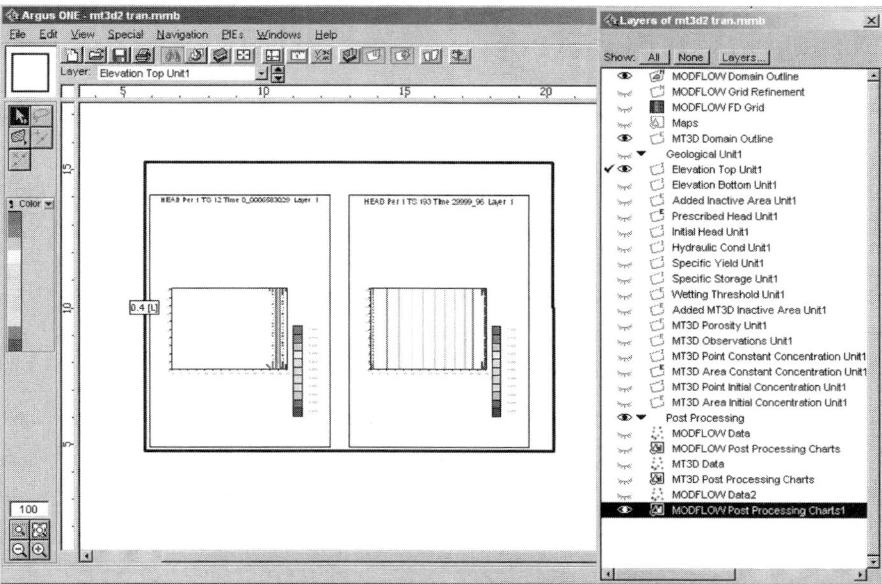

FIGURE 3.40. *Contour plot of the head values calculated by MODFLOW.*

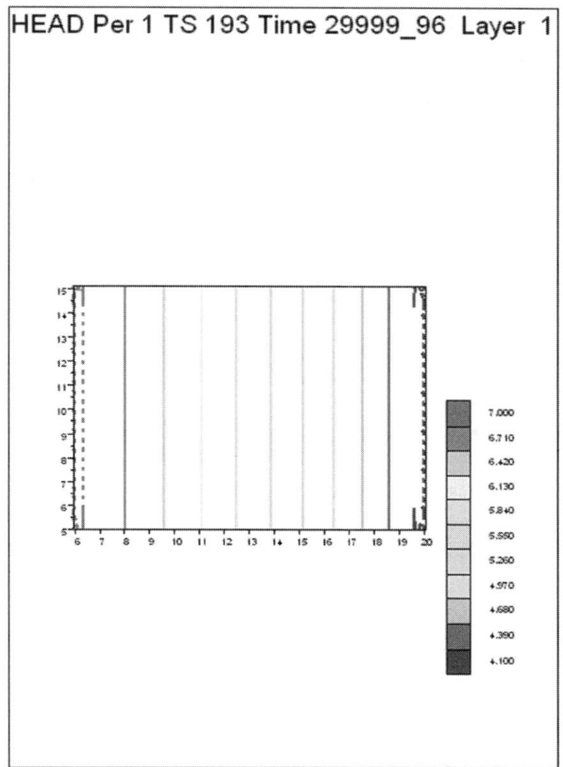

FIGURE 3.41. *Head contours calculated by MODFLOW at the end of time step 193.*

contours at the end of time step 193. As expected, given specified, y-invariant head values at each end of the domain, the flow is uniformly left to right and the contours are approximately equally spaced.

PTC Postprocessing The postprocessing protocol for flow solutions using *PTC* is similar to, but different than, that used by *MODFLOW*. In the *PTC Configuration* dialog window (see Figure 3.3) we specified the file prefix to be used for the files containing the head results generated by *PTC*. We selected the letter *s*.

Access to the results is achieved by activating the *PTC Data* layer, selecting *File*, then *Import PTC Data...*, and finally, *Text File*. The *Import Data* dialog window opens and it is necessary to change the default by pushing the *Mesh data* and *Read triangulation from layer* radio buttons. Clicking *OK* brings up the *Choose file to import* dialog box. The default value appearing in the *Files of type:* drop-down list must be changed to *All Files (*.*)* to reveal all of the files available in the *PTC-Output* folder. The files used directly by the executed *PTC* run are those found in the

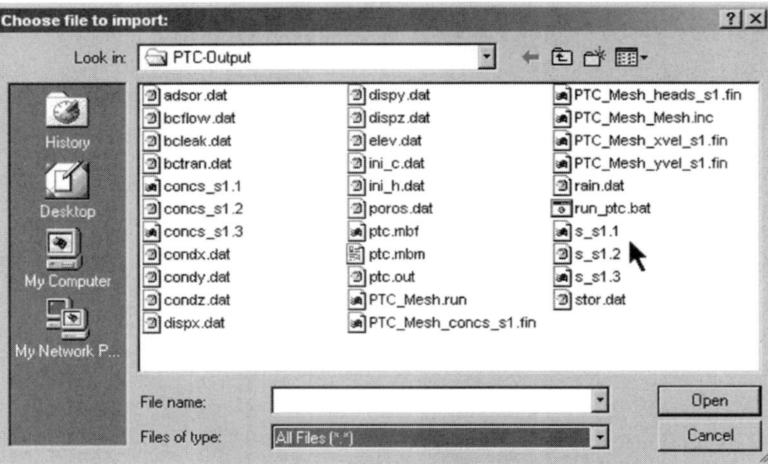

FIGURE 3.42. *The Choose file to import dialog box from which one selects the files to be plotted.*

list box in Figure 3.42. Those of interest to us are indicated by the location of the pointer. Each represents one output file of calculated head values.

To determine the time associated with each of these files, one must interpret the *PTC* configuration file presented in Figure 3.43. The total number or time steps is

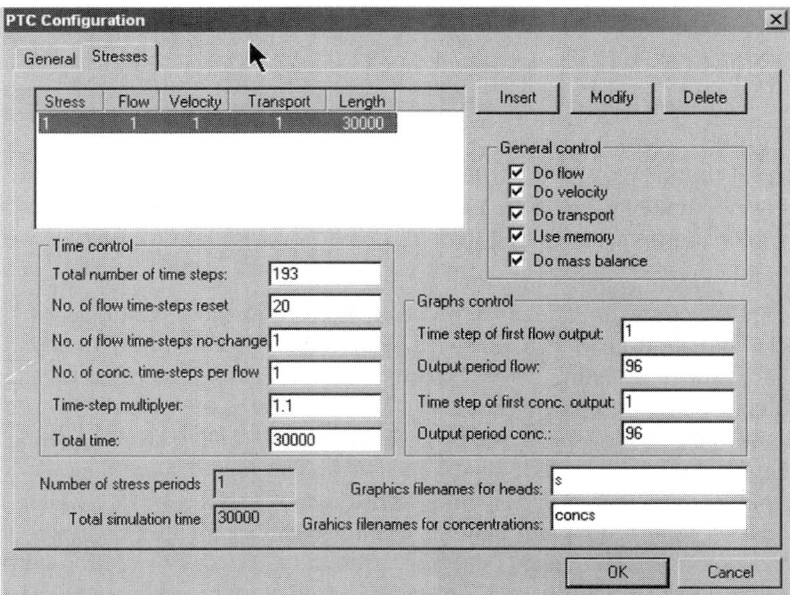

FIGURE 3.43. *PTC Configuration file that provides the information needed to select the output files from the Choose file to import dialog window.*

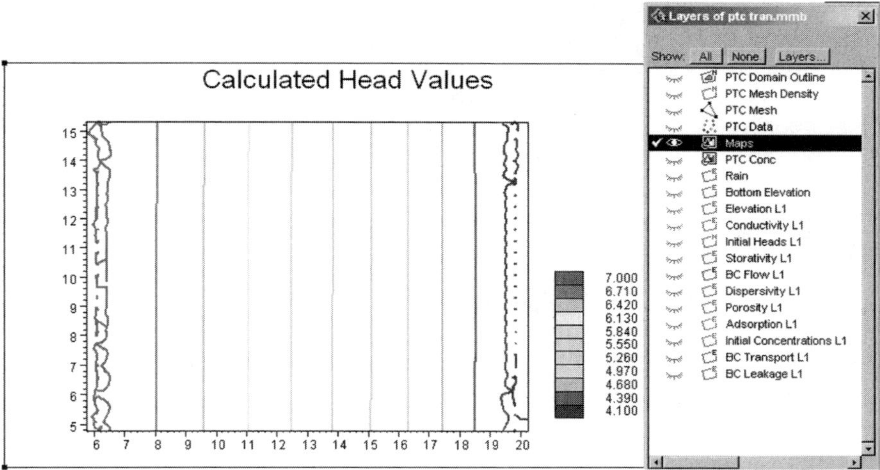

FIGURE 3.44. *Contour plot of final head values obtained using PTC.*

193. The first output is at step one. The next will then be at step 97 and the last at 193. Thus, file *s_s1.1* is the solution for the first time step and *s_s1.3* is the solution at the last time step. Calculation of the elapsed time associated with each of these two steps is easy. However, determining the time elapsed for step 97 is a challenge.

The graphical output is generated by activating the *Maps* layer. A set of tools will appear in the upper left-hand corner of the screen. The tool in the lower right-hand corner of the array (it looks like a sheet of paper) is the *Post-processing* tool and is used to select the form of the output. Six possible output vehicles are available by clicking and holding the tool. We have selected the contouring option (target-shaped icon) in presenting the results shown in Figure 3.44. The several intervening steps required to obtain the output are described in Section 1.17.

3.1.2 Groundwater Transport

Initial Data Unlike *PTC*, groundwater transport simulation using the flow results obtained using the basic *MODFLOW* simulator can be realized with more than one software package. In this section we use *MT3D Basic. MOC3D* is also a popular option. The *MODFLOW/MOC3D Data Sets* dialog box is shown in Figure 3.45. Checking the *Use MT3D* box indicates that *MT3D* is the transport simulator of choice. In the *MT3D Heading Line* text boxes the title to be associated with the transport output is specified. In the *Concentration applied to inactive concentration cells* text box, a value is provided such that solutions at the inactive cells will be readily identified in the transport output. The purpose of this information is to assist in delineating those cells that are intended to be external to the area wherein concentration values are to be calculated and plotted. The *Length Unit* and *Mass Unit* text boxes are used to specify the units to be used for length and mass, respectively. The *Sink and Source Mising* box is checked if prescribed heads with nonzero concentrations or

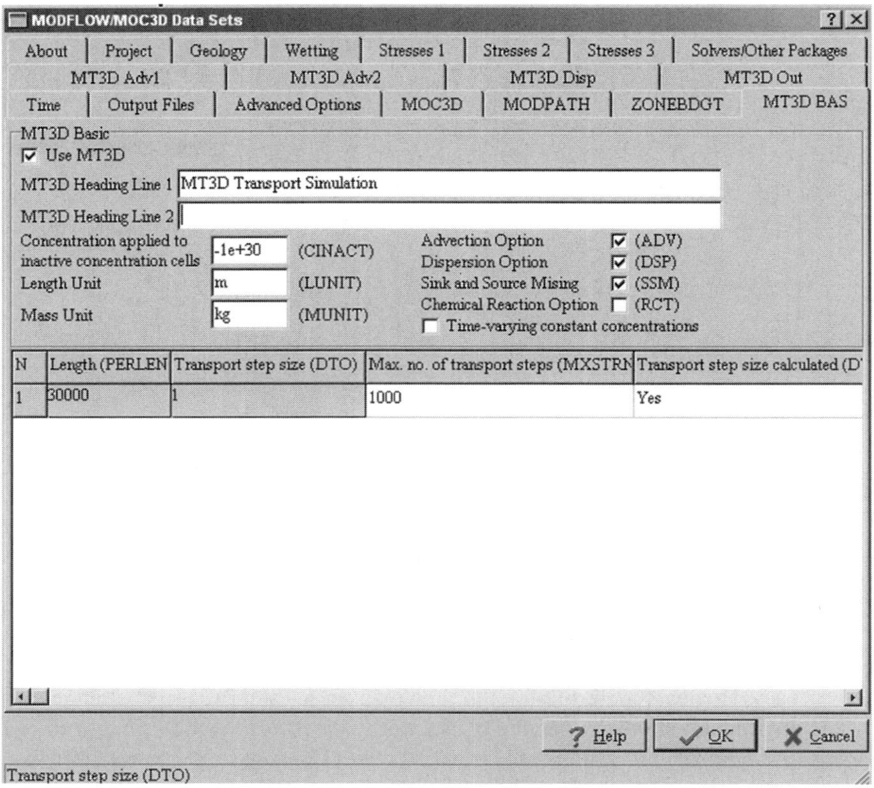

FIGURE 3.45. *MODFLOW/MOC3D dialog box used to set general parameters for the transport simulation.*

time-varying constant concentrations are to be considered in the model. To use the chemical-reaction package, the *Chemical Reaction Option* box must be checked. No chemical reactions are considered in this example, so this box remains unchecked. If the concentrations specified at source locations are time varying, the *Time-varying constant concentrations* box is to be checked.

The *Advection Option* and *Dispersion Option* text boxes are checked if advective- and dispersive-transport mechanisms are to be considered in the simulation. In our example both of these mechanisms are considered.

The large list box at the bottom of the screen shown in Figure 3.45 is used to provide information on the time-step size to be used for each computational period *N*. The length of period *N* is given by the value found in the *PERLEN* text box and is provided by activation of the *Time* tab that is used to specify the length of the *MODFLOW* simulation. In the *Maximum no. of transport steps* text box, place the number of steps you wish to take in order to simulate the length of period *N*. The information on the period length and the number of steps is used to calculate the

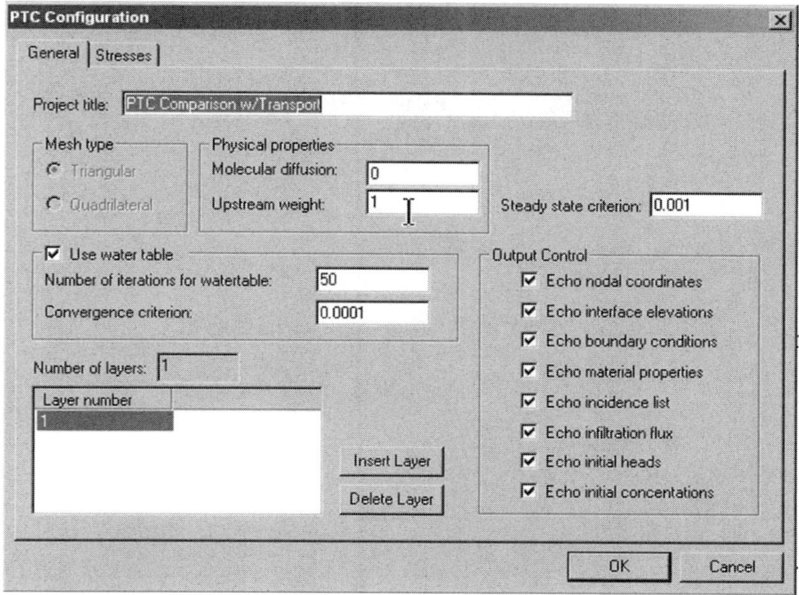

FIGURE 3.46. The information needed to activate a transport simulation using PTC requires specification of Molecular diffusion and Upstream weight. Greater efficiency is achieved if a Steady state criterion is also provided.

size of each time step. If the user wishes to have the time-step length calculated automatically by *MT3D*, toggle to *yes* in the *Transport step size calculated* text box. Note that in our example, we elect to have time steps calculated automatically.

The initial information required to run the transport option in *PTC* is minimal. Activating the *Stresses* tab in the *PTC Configuration* dialog window reveals the screen shown in Figure 3.46. The only information to be indicated here are the *Molecular diffusion* and the *Upstream weight*. The upstream weight is unity for complete upstream weighting and 0.5 for no upstream weighting. Upstream weighting is discussed in more detail in Section 2.6.

Activation of the *Stresses* tab reveals the window shown in Figure 3.47. Note that the *Do transport* check box is activated. In addition, several *Time Control* and *Graphs control* options associated with transport must be considered. The ability to calculate several transport time steps per flow time step is provided by the *No. of conc. timesteps per flow* dialog box. In our example we use the same-size time step for both transport and flow and therefore we have selected a value of *1* for this parameter. As in the case of flow, one can specify when the first transport output should take place and the time-step interval between outputs. For example, in Figure 3.47 there will be 96 time steps between output of concentration values. One must also specify the name that is to be used as part of the concentration output file names.

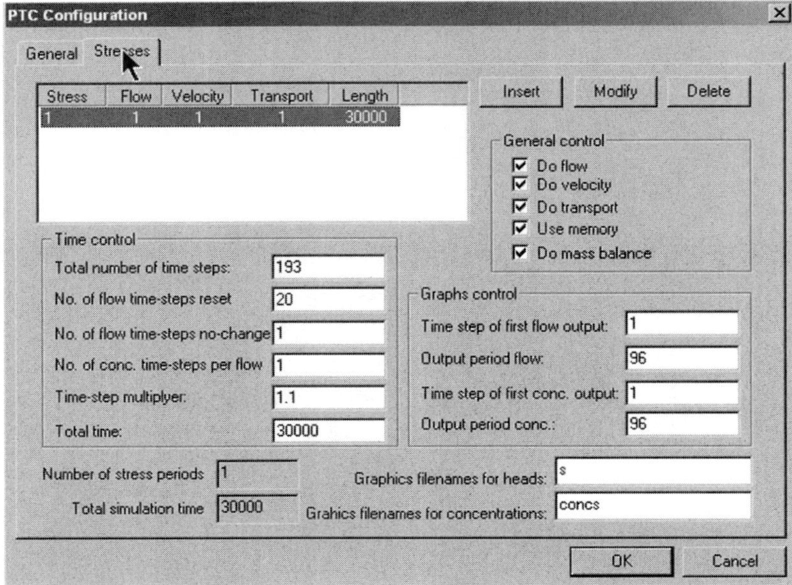

FIGURE 3.47. Dialog window revealed by activating the Stresses tab of the PTC Configuration window. This window is now configured for transport.

Model Geometry The model geometry for transport is assumed to be the same as that for flow in the case of *PTC*. Thus if there are areas that should exhibit mesh densities for transport that differ from that required for flow, those densities must be incorporated into the single mesh-density specification.

In the case of *MT3D*, it is possible to modify the region to be considered for flow to accommodate the transport calculations more efficiently. The two vehicles to be used for that purpose are the *MT3D Domain Outline* and the *Added MT3D Inactive Area Unit1* layers. Neither of these options are used in our example problem.

Boundary Conditions Having activated *MT3D* in the *MODFLOW/MOC3D Data Sets* dialog window, we now turn our attention to specification of the concentration-based transport boundary conditions. In Figure 3.48 one observes that one has the option of defining a point or an area concentration boundary condition. In the case of our example, the highlighted *MT3D Area Constant Concentration Unit1* condition is specified. The area specified to have a concentration of 1000 mg/cm^3 is indicated in the area to the left of the layers drop-down list. The specified-concentration conditions are defined in a manner analogous to that used in identifying specified-head conditions. Addition of a second-type or constant diffusive mass-flux boundary condition can be accomplished using the well package (*Stresses 1* tab). One can then specify the desired flux concentration at a point location. By default, around the perimeter of the region, a no-diffusive (or dispersive) mass-flux condition is assumed.

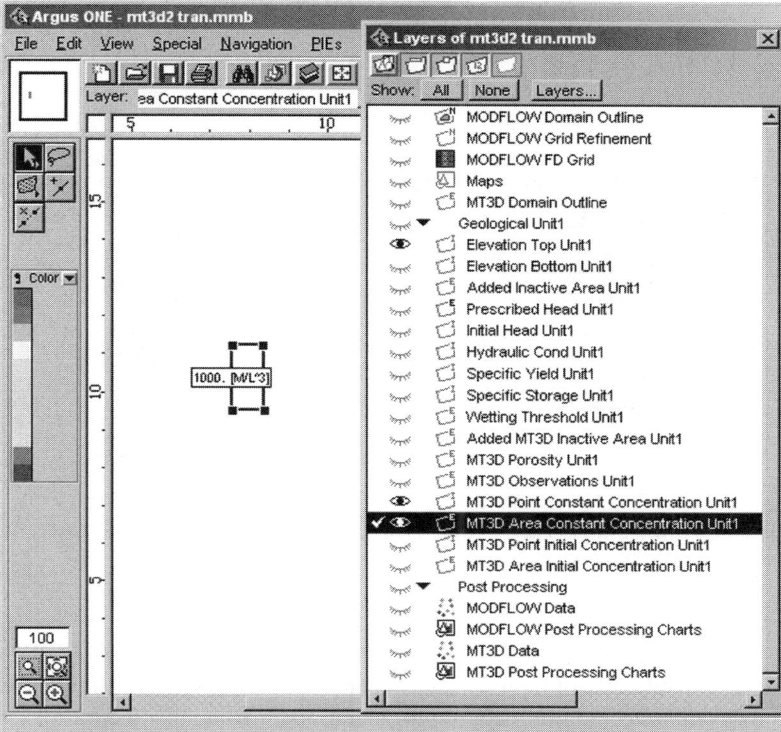

FIGURE 3.48. *Layers provided by MT3D for specification of concentration boundary conditions and the specific condition appropriate to our example problem.*

Let us now turn our attention to the specification of concentration boundary conditions using *PTC*. As in the case of *MT3D*, the area concentration conditions are specified using the contour tool shown in Figure 3.49. In the case of *PTC*, the modeler has the option of defining both specified concentration (type 1) and specified diffusive mass flux (type 2) conditions using *BCTransport L1*. As in the case of *MT3D*, an undefined external boundary is assumed to be a no-diffusive (or dispersive) mass-flux condition. A point-concentration boundary is specified using the point-contour tool in a manner analogous to that used to define a well in the case of groundwater flow.

Initial Conditions In the case of either *MT3D* or *PTC*, initial conditions are specified in the same manner as any of the earlier field parameters, such as initial head. The appropriate protocol is given in Section 3.1.1. In the case of this and most contaminant transport problems, the initial condition on concentration is given as zero. The reason for this is that most transport simulations start from pristine aquifer conditions.

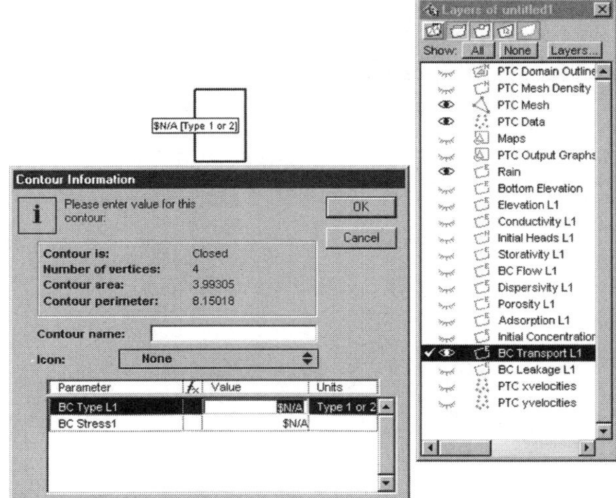

FIGURE 3.49. *Layers provided by PTC for specification of concentration boundary conditions and the specific condition appropriate to our example problem.*

If nonzero initial conditions are specified, one must be aware of the fact that these conditions probably do not represent mass-conservative conditions. In other words, using observed concentration data appropriately extrapolated to the various nodal locations will normally result in a computed redistribution of mass as a natural consequence of the model attempting to provide a mass-conservative concentration solution.

Spatial-Data Input Other than the initial-concentration conditions specified in Section 3.1.2, the only other standard transport parameter not specified earlier in defining the input for the flow calculation is the dispersivity (see Section 2.4.2). In the case of *MT3D*, this information is specified via the *Hydraulic Cond Unit1* layer (see Figure 3.23) or via a special package to which access is provided in the *MODFLOW/MT3D Data Sets* dialog window (Figure 3.45). In the case of *PTC*, the information is specified as a distributed field parameter via the *Dispersivity* layer (Figure 3.50).

This is somewhat unusual since it is customary to group transport-related parameters together and not intermix them with flow parameters. The cursor in Figure 3.23 identifies the location where the *Logitudinal Dispersivity* is placed. One should take notice of the fact that the units of dispersivity are length, not length over time as in the case of hydraulic conductivity.

Model Execution Because transport simulation is but an option in *PTC*, the only action needed to initiate a run is to activate the standard run sequence outlined in Section 3.1.1. Prior to the launch of *MT3D*, one must already have run *MODFLOW*

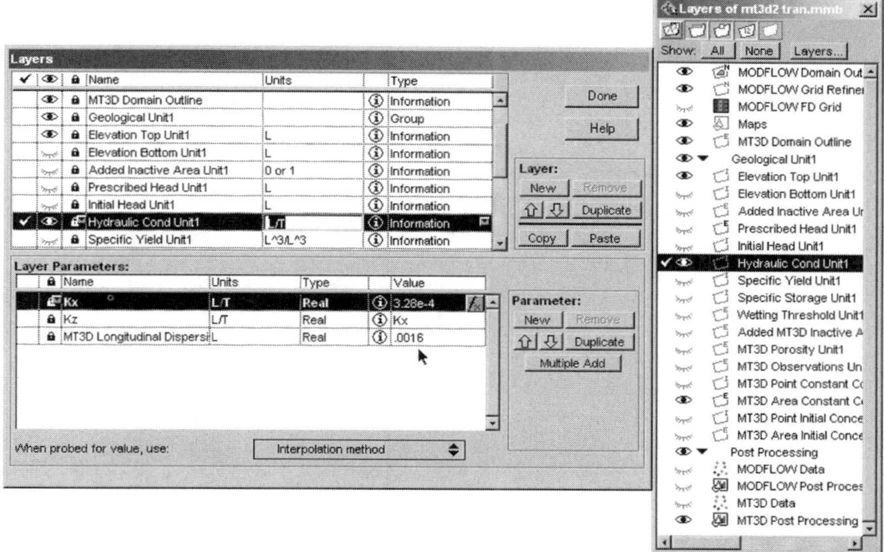

FIGURE 3.50. *Dialog box and drop-down list used to define the input of hydraulic conductivity and Dispersivity in MODFLOW.*

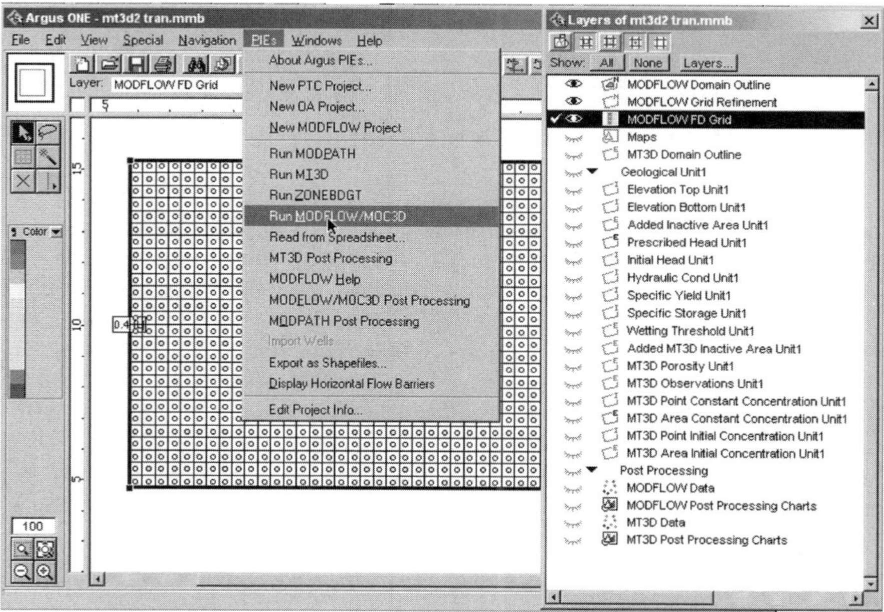

FIGURE 3.51. *To launch MT3D, one first selects the MODFLOW/GRID layer and then PIEs option. From the drop-down window select the Run MT3D option.*

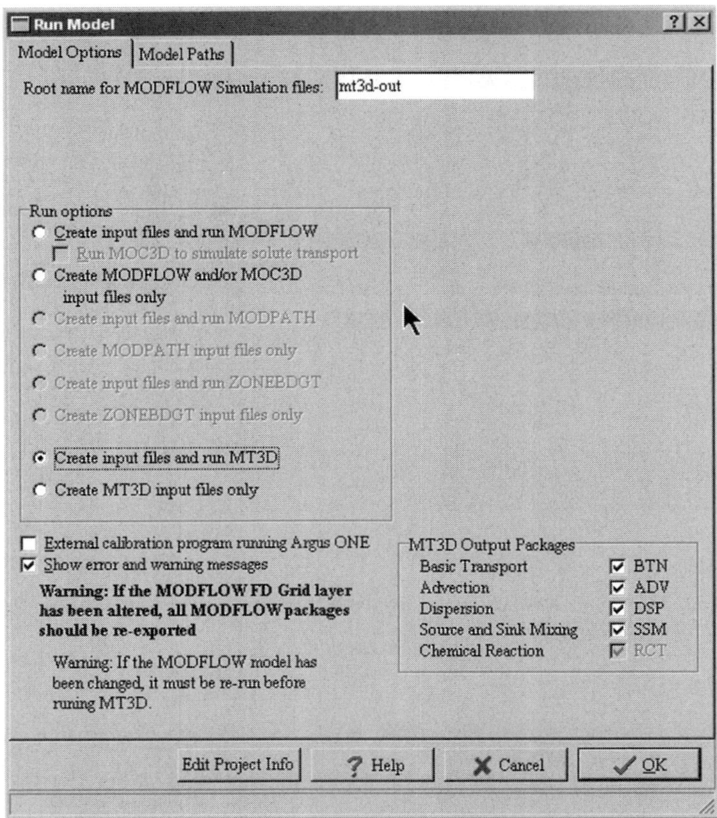

FIGURE 3.52. *Window activated by selecting the run option for MT3D.*

to generate the necessary flow-field information. If this is not done, on the attempted launch of *MT3D*, an error message indicating the missing files will be reported.

To begin the *MT3D* run, one first activates the *Modflow Grid* layer (Figure 3.51). Next, select *PIEs* from the main *Argus ONE* window and then select either *Run MT3D* or *Run MODFLOW/MOC3D* from the drop-down list (Figure 3.51). The window shown in Figure 3.52 appears. Push the radio button associated with the *Create input files and run MT3D* option. A window informing the user that the *Processing Export Template* is active now appears. Upon completion of the export, a list box appears requesting the file name for the export file. An *Export in Progress* window analogous to that generated for *PTC* now appears. When the export is complete, *MT3D* is launched and the window shown in Figure 3.53 documents the progress of the simulation.

Postprocessing of Transport Output The postprocessing protocol of the *PTC* concentration values is identical to that used to view the computed head values. The

```
C:\WINDOWS\System32\cmd.exe
TIME STEP NO.  191
FROM TIME =    22539.      TO     24793.
Transport Step:     3    Step Size:    429.0      Total Elapsed Time:    24793.
TIME STEP NO.  192
FROM TIME =    24793.      TO     27273.
Transport Step:     3    Step Size:    653.6      Total Elapsed Time:    27273.
TIME STEP NO.  193
FROM TIME =    27273.      TO     30000.
Transport Step:     3    Step Size:    901.2      Total Elapsed Time:    30000.
C:\PROGRA~1\Argus\ArgusPIE>Pause
```

FIGURE 3.53. *Window generated by MT3D on activation of the simulator. The provided information is the time step number, the time at the beginning of the time step, the time at the end of the time step, the time step size, and the transport step.*

only difference is the selection of the concentration rather than head files when the window appearing in Figure 3.42 appears. The concentration contours created using this protocol for a period of 30,000 time units is presented in Figure 3.54.

Postprocessing of the *MT3D* data is achieved by first selecting the *MT3D Post Processing* option that appears in the drop-down window when the *PIEs* option is selected from the *Argus ONE* main window. The resulting window is given in Figure 3.55. Of importance to us is activation of the *MT3D* radio button and *No Transformation* button. Default values of the other parameters are appropriate for our example problem. Clicking the *Import Data* button reveals a dialog window from which one selects the *.UCN* concentration files to be considered for plotting. Closing

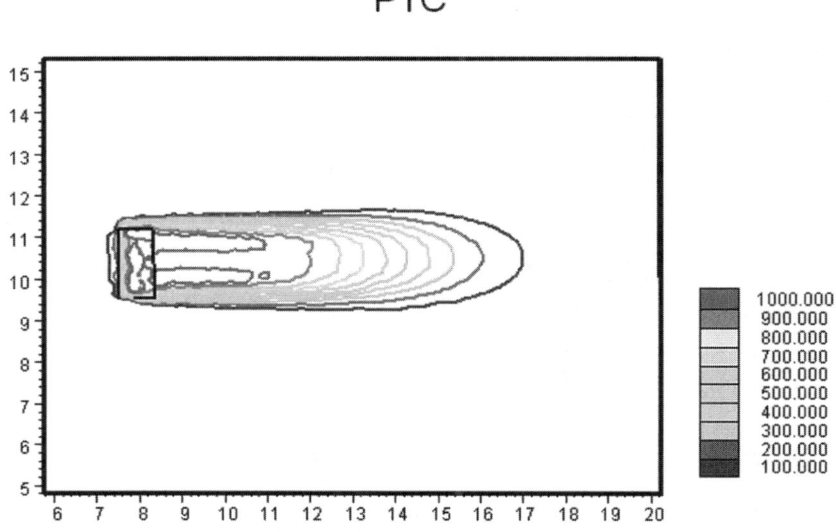

FIGURE 3.54. *Concentration contours generated by PTC for a time of 30,000 time units.*

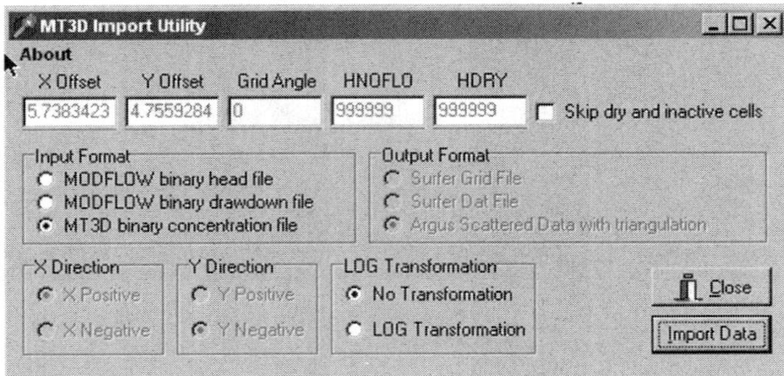

FIGURE 3.55. MT3D-generated window wherein one designates the nature of the proposed plot. In this case the MT3D binary concentration file is selected and there is to be no transformation of the raw data to Log format.

this window activates another dialog window, from which one selects the *.CNF* grid configuration files to be used in the plotting procedure.

Upon selection of the concentration file to be considered for plotting, the *Output Format* window shown in Figure 3.56 is presented. Various options are provided for presentation of the simulation results. We are interested in an areal view, so the *Layer* radio button is selected. Since the finite-difference mesh consists of rectangles, the area to be plotted is specified by selecting the appropriate rows and column of the mesh. In this case columns 1 through 36 and rows 1 through 26 are considered.

Clicking OK in the *Output Format* window activates the *Choose Time Step* window shown in Figure 3.57. We are interested in viewing the last time step, which represents the solution at an elapsed time of 30,000 time units, and thus the last available box is the one checked. If the file you wish to plot has already been selected, a dialog window opens to allow the user to specify whether to overwrite the existing file or change the name of the file.

The final step is to select the type of chart to be provided. The options are revealed (Figure 3.58) when the *OK* button in the *Choose Time Step* window is pushed. This is analogous to using the *Postprocessing* tool in the *PTC* output protocol. Selection of

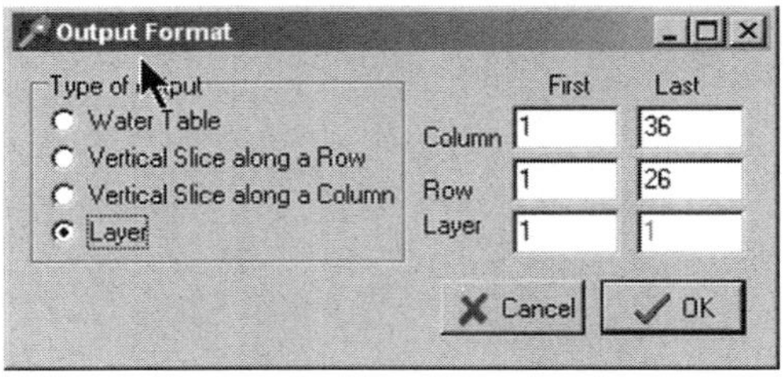

FIGURE 3.56. MT3D Output Format window wherein one selects the kind of plot to be produced. In our example, an areal view of one layer is requested by selecting the Layer radio button.

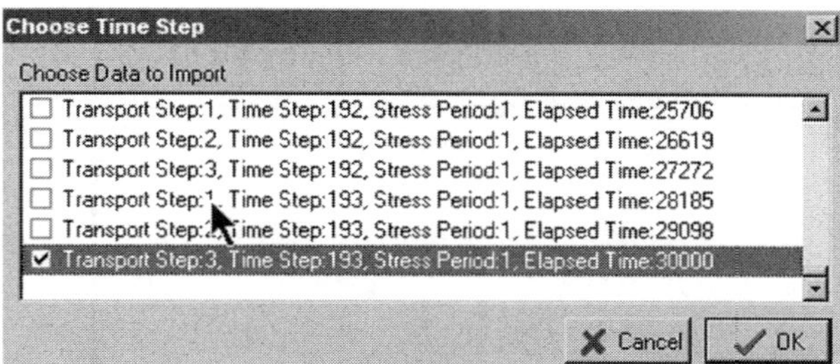

FIGURE 3.57. *The Choose Time Step window is used to select the time steps for which graphical output is desired.*

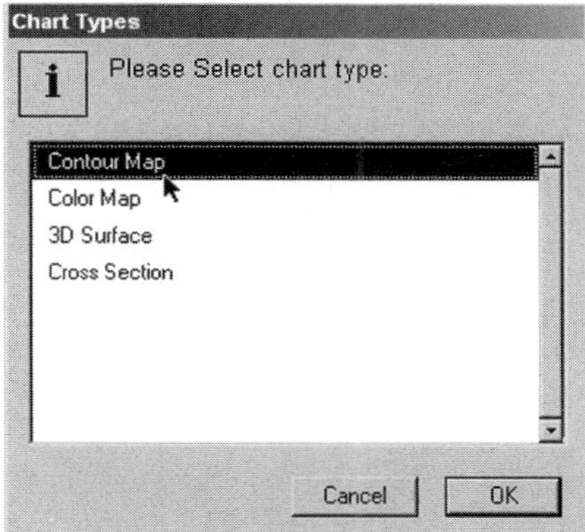

FIGURE 3.58. *The MT3D output format is selected using this window.*

Contour Map activates the plotting program, which generates the graphic presented in Figure 3.59. The results presented in this figure can be compared with those provided in Figure 3.54.

3.2 COMPARISON OF METHODS

Examination of the preceding sections of this part reveals that the differences between the finite-element and finite-difference attributes can be subdivided into those associated with the graphical user interfaces (GUIs) and those attributable to the modeling techniques.

3.2.1 Graphical User Interfaces

Let us first consider the attributes of the GUIs. *MODFLOW*, *PTC*, and *MT3D* use the Argus ONE numerical environment. The difference, in application, can be attributed to the plug-in extensions (PIEs) developed for each. A review of the preceding sec-

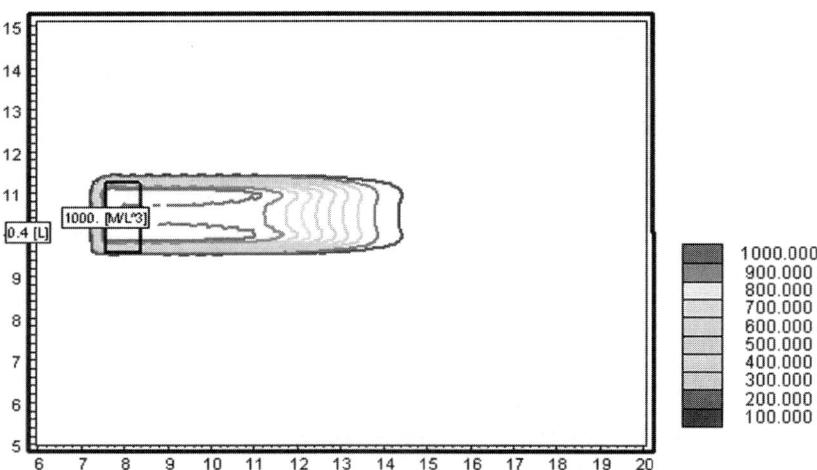

FIGURE 3.59. *Concentration contours generated by MT3D.*

tions of this part reveals that although the two PIEs both provide input to groundwater flow and transport models using the Argus ONE numerical environmental, they are quite different.

The *MODFLOW* and *MT3D* PIEs are designed for the user who benefits from step-by-step guidance in model development. The dialog boxes are designed to accommodate the needs of those who are accustomed to an interface that facilitates the practical application of modeling. The various specialty tabs found in the main *MODFLOW/MOC3D* window (see Figure 3.45) attest to the effort dedicated to facilitating the use of the *MODFLOW* and *MOC3D* computer codes by the less experienced user. In particular, ready access is provided to numerous support packages. The *Help* tab that is accessible in many of the windows is particularly useful.

The *PTC* GUI is more cryptic and assumes a more mathematically sophisticated user than is the case for *MODFLOW*. It is assumed, for example, that the analyst is comfortable translating hydrological conditions into mathematical statements. Although at first blush this may seem like a disadvantage, such may not be the case. The more experienced modeler may prefer to have more control over the mathematical interpretation of the physical attributes of the system to be modeled. In addition, there is a practical trade-off between the efficiency of a concise, albeit more abstruse mathematical statement of a physical problem and a more transparent, user-friendly, but possibly less precise alternative.

3.2.2 Model Formulation and Implementation

MODFLOW-MT3D The primary differences between the *MODFLOW-MT3D* model package and the *PTC* finite-element program are the program structure, the assumptions that can be made regarding the flow in low-conductivity layers, the treatment of water-table conditions, and use of a rectangular versus a triangular mesh. *MODFLOW* is a finite-difference model that is used to solve groundwater-flow equations. It is used to provide the groundwater-velocity information needed to solve the *MOC3D* and *MT3D* transport models. Inasmuch as it is finite-difference based, it generates systems of equations that are amenable to solution by a number of efficient solvers, including those that are based on iterative-solution methods. As currently formulated, the three-dimensional version of the program has the option of treating flow in low-hydraulic-conductivity layers as vertically one-dimensional. Treatment of the water-table option is pragmatic if somewhat heuristic and care must be taken in its application.

The major feature that separates finite-difference models such as *MODFLOW* from finite-element models such as *PTC* is the mesh geometry. Finite-difference models generally require use of a rectangular mesh. The practical consequences of this limitation as implemented in *MODFLOW* are threefold:

1. Once initiated, a mesh line can be terminated only at the model boundaries.
2. The distance between two mesh lines must remain constant.
3. Model boundaries must be represented by rectangular blocks.

The practical ramification of observations 1 and 2 is that it is awkward to introduce mesh refinements. If a small mesh block is introduced, the lines associated with this block must extend to the model boundaries, with the potential of generating long, narrow elements that exhibit extreme aspect ratios. Extreme aspect ratios may lead to challenges using some algebraic equation solvers. Care must also be taken in mesh design to avoid large changes in adjacent-element geometry. In other words, one should not allow Δx, Δy, or Δz to change too rapidly from node to node. A rule of thumb is to require that $1 \leq \Delta x_{i+1}/\Delta x_i \leq 1.5$. In Figure 3.60 the area to which

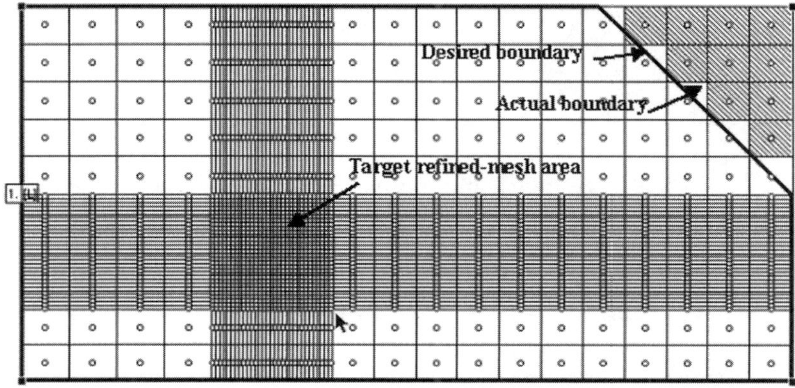

FIGURE 3.60. *MODFLOW grid generated to illustrate boundary approximations and refinement.*

the cursor is pointing has been enhanced to accommodate a well. In this area, Δx and Δy have been reduced by a factor of 10. Note that the refined grid extends to the model boundaries and that the elements outside the area of desired refinement have an aspect ratio of 10. If one considers the coefficients in the matrix equation generated by the finite-difference approximation to the flow equation, one finds that those elements containing approximations of the form $(\cdot)/(\Delta x)^2$ will have a magnitude approximately 100 times smaller than those those of the form $(\cdot) / (\Delta y)^2$ in regions to the east and west of the target area for refinement. In the areas to the north and south of the target area, terms of the form $(\cdot) / (\Delta y)^2$ are approximately 100 times smaller than those of the form $(\cdot)/(\Delta x)^2$. While modern computers can handle matrices with this degree of ill-conditioning, it is clear that care must be exercised in finite-difference grid refinement on a rectangular mesh to assure that the aspect ratios do not get out of hand.

In the northeast corner of the model area shown in Figure 3.60, use of rectangles to define a nonrectangular boundary is illustrated. The boundary, as viewed from the point of view of the finite-difference model, is the area shown in white. The boundary itself is defined by the line separating the white and patterned areas. The desired boundary is the bold diagonal line running from the northwest to the southeast.

PTC *PTC* is a hybrid code using a finite-difference approximation in the vertical plane and a finite-element formulation in the horizontal plane. Thus each three-dimensional element is a column with triangular cross section. A splitting algorithm is used to approximate the flow and transport equations. The horizontal layers are solved first and then the vertical lines. The result is a fully three-dimensional solution that is achieved without having to solve a matrix generated from a three-dimensional mesh.

Because a finite-element mesh is used in the horizontal plane, the limitations presented in observations 1, 2, and 3 are not applicable. For example, the finite-element mesh generated for the problem presented in Figure 3.60 is given in Figure 3.61. The

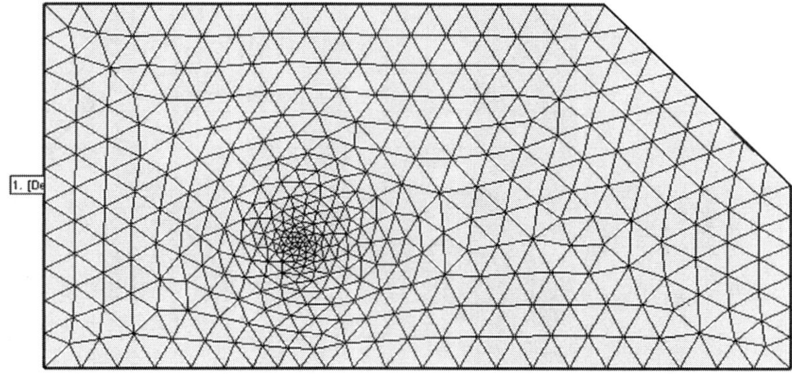

FIGURE 3.61. *Finite element mesh automatically generated for cut-corner example problem.*

density in the neighborhood of the well is the same as that presented in Figure 3.60. The number of nodes in the area around the well is much smaller, although the exact number is a function of the area around the well defined in the *MODFLOW* grid specification. The total number of nodes in the finite-element mesh is approximately 400, and in the finite-difference mesh, about 1600. However, the number of nodes in the finite-difference mesh is a function of the area selected for grid refinement. A smaller area of grid refinement would generate a smaller total nodal number.

The northeast boundary of the model (i.e., the one defined by the diagonal boundary), is faithfully reproduced by the finite-element mesh. Although the number of nodes in the finite-element mesh is approximately one-fourth of the number in the finite-difference mesh, the computational effort is not 75% less. The finite-element mesh has less structure, which translates into a less structured coefficient matrix. Indeed, our limited experiments have indicated that the amount of computational effort required to solve a practical problem is approximately the same whether a finite-difference or a finite-element formulation is used.

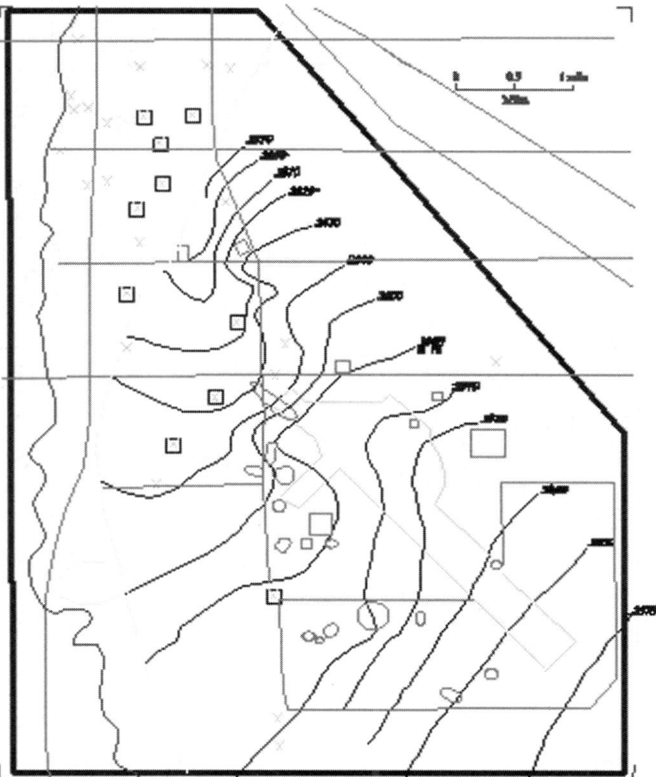

FIGURE 3.62. *Tuscon area of investigation for a comparison of the finite-difference and finite-element methods.*

Next we examine a few example problems solved using the finite-element and finite-difference methods.

3.2.3 Groundwater Flow

The practical differences between the finite-difference and finite-element methods is perhaps best illustrated using the Tucson example presented in Figure 3.62. The contours shown are the water levels observed. Figure 3.63 illustrates the finite-difference grid used to solve this problem. There are approximately 1100 nodes in this mesh.

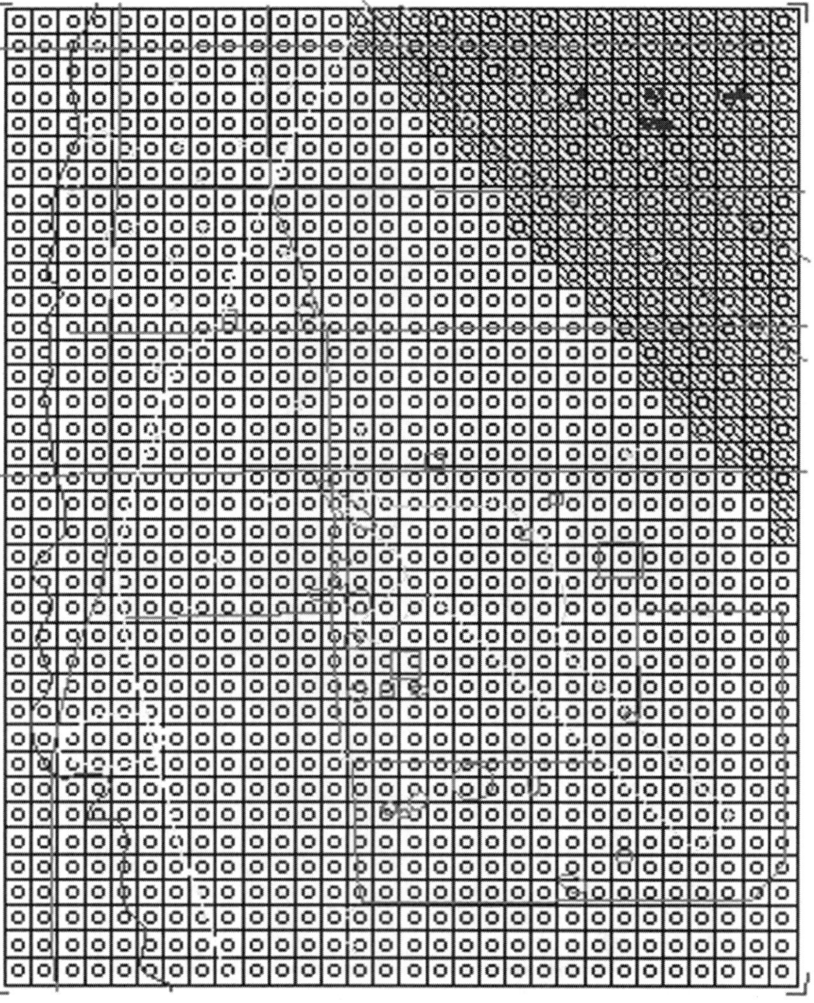

FIGURE 3.63. *Finite-difference grid used to solve the Tuscon problem.*

The following observations are worthy of note:

1. The mesh is uniform and is not automatically refined in the neighborhood of singular points such as wells. It is possible, as shown earlier, to refine the mesh as needed as a separate step.
2. The northeast boundary is represented by a steplike boundary made up of the perimeters of rectangular finite-difference cells.

A transient solution obtained using this formulation is shown in Figure 3.64. The computed and observed hydraulic-head contours differ, but the difference is typical of model results that are obtained in a preliminary analysis.

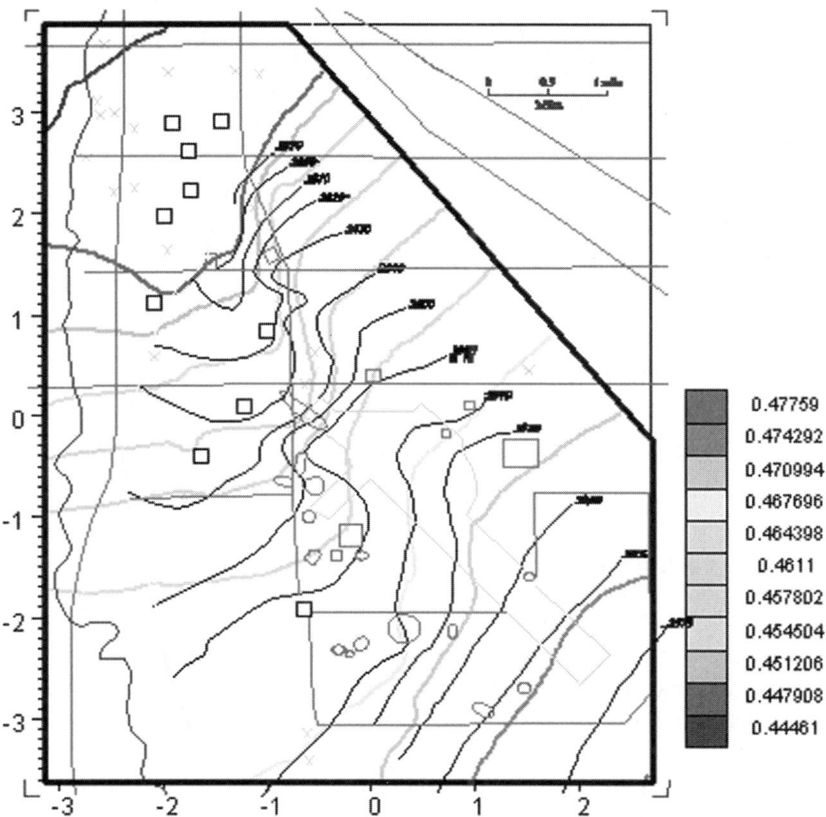

FIGURE 3.64. Transient finite-difference solution of the hydraulic head at Tucson Arizona site obtained using the formulation shown in Fig. 3.63.

PTC Transient

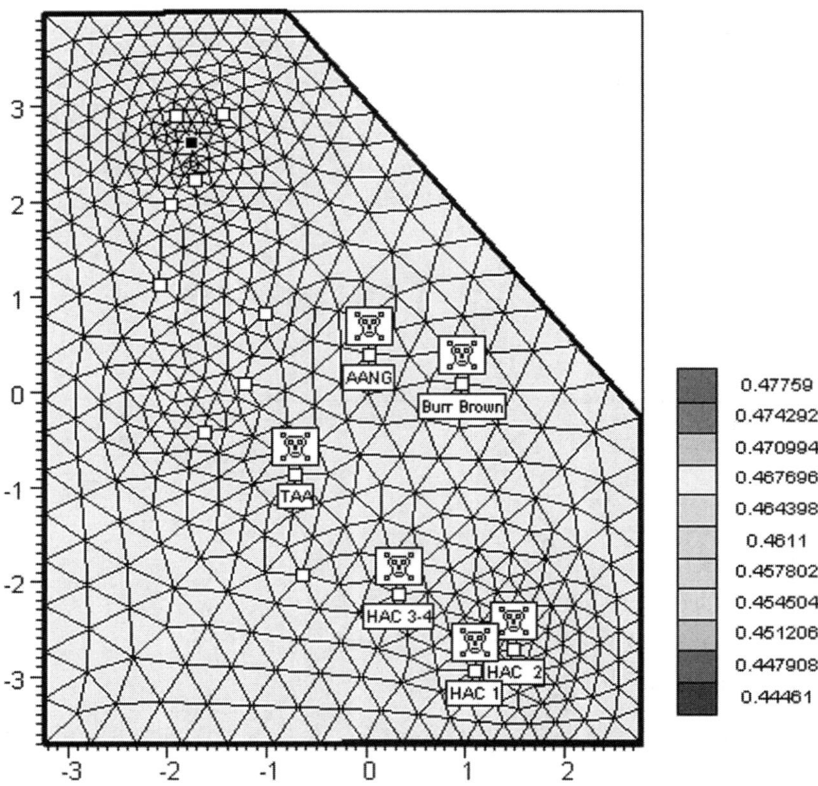

FIGURE 3.65. *Finite-element mesh used to solve the groundwater-flow problem at Tucson, Arizona.*

The same problem has been studied using the finite-element formulation. The mesh used, shown in Figure 3.65, contains 500 nodes. The following observations are noted:

1. The number of nodes and elements in the finite-element mesh are approximately half those obtained using the finite-difference approach (see Figure 3.63).
2. The mesh is automatically refined in the neighborhood of singular points such as pumping wells, where the hydraulic-head gradient is very steep.
3. The boundary in the northeast corner of the model is represented exactly by the finite-element mesh.

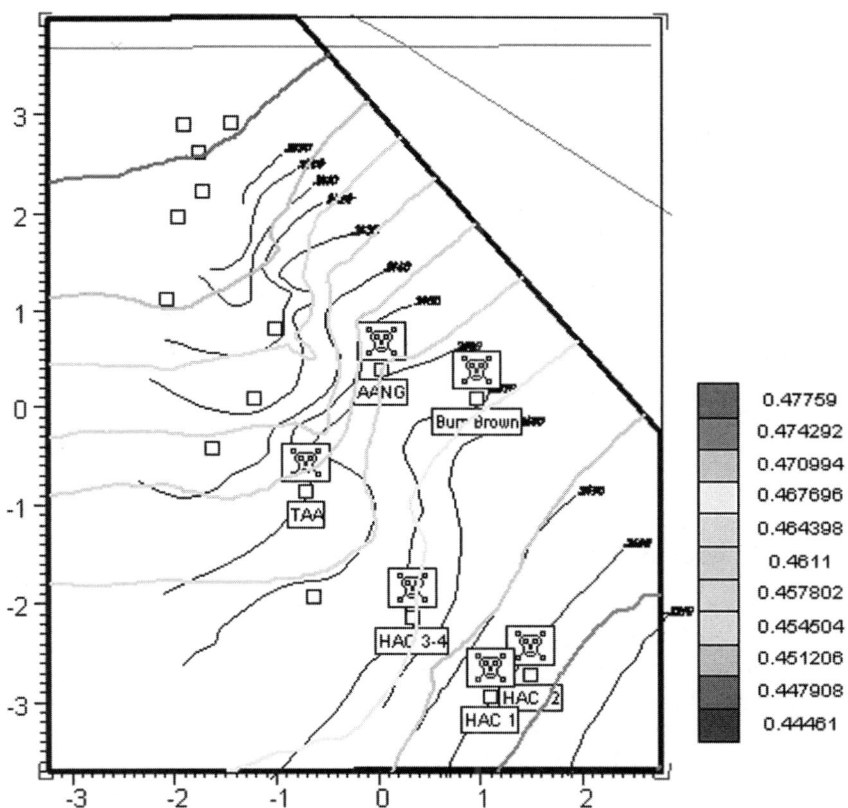

FIGURE 3.66. *Transient solution for hydraulic head obtained using the finite-element formulation shown in Fig. 3.65.*

The transient solution for the hydraulic head obtained using the finite-element formulation shown in Figure 3.65 is presented in Figure 3.66. Let us now compare the solutions obtained using the finite-difference and finite-element formulations. Both models reproduce the observed contours to approximately the same degree, although the areas of optimal fit are different in each model. The boundary conditions, especially the no-flow condition along the northeast boundary, are accommodated accurately (i.e., the contours are perpendicular to the no-flow boundary). The *PTC* contours appear to respect the presence of wells more accurately (i.e., the contours are deformed more appropriately to indicate the presence of the wells). This is as expected since the finite-element mesh is more refined in the area of the wells.

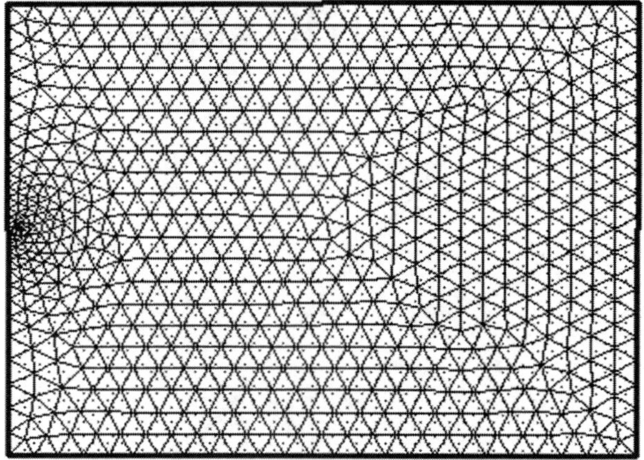

FIGURE 3.67. *Finite element mesh used to obtain the concentration solution shown in Fig. 3.68.*

3.2.4 Groundwater Transport

Groundwater transport is a much more challenging problem to model than is groundwater flow. The addition of the convective term, which translates into a spatial first derivative in the governing partial-differential equation, is notoriously difficult to handle numerically. In essence, the challenge is met by having a very refined mesh in areas where the concentration gradient is steep. To illustrate the importance of this concept, we conduct a numerical experiment. Consider the problem presented in Figures 3.54 and 3.59. A standard finite-element grid (i.e., one with no enhanced refinement) is shown in Figure 3.67. Because the source area is a point boundary condition, the mesh is automatically refined in this area. The concentration solution generated using this mesh is shown in Figure 3.68.

In Figure 3.69 the same basic mesh has been refined in the neighborhood of the source. Because the mesh is so fine in this area, it is presented in magnified form in Figure 3.70. The refinement illustrated here is extreme and is presented only to demonstrate the importance of mesh refinement on transport-solution accuracy.

The solution obtained using the grid shown in Figure 3.69 is presented in Figure 3.71. The solution is for the same conditions and elapsed time as shown in Figure 3.68. The solutions are dramatically different. The 0.2 contour (second in from the outside boundary), for example, has moved much closer to the source. In general, the contours are more closely spaced in the neighborhood of the source in Figure 3.71 than in Figure 3.68. This is the practical ramification of numerical dispersion, a phenomenon attributable to the use of large grid spacing.

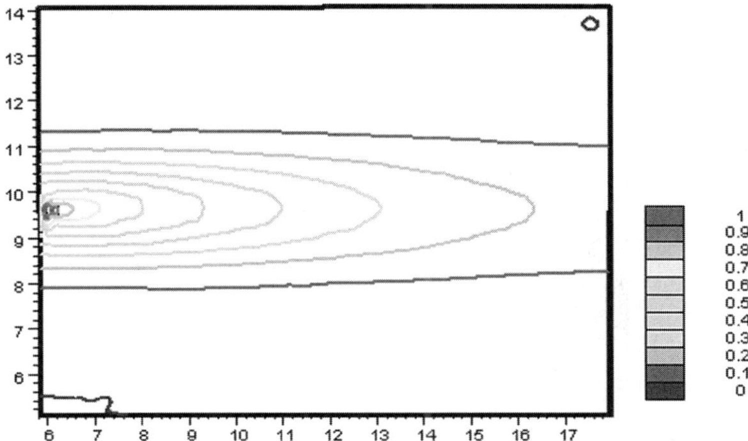

FIGURE 3.68. *Finite-element concentration solution obtained using the grid shown in Fig. 3.67.*

3.3 SUMMARY

A comparison of finite-difference and finite-element models of groundwater flow and transport created via the Argus ONE numerical environment reveal a number of basic similarities and differences. The methods are similar in the following ways:

1. Both modeling methods use an approximation for the behavior of the water table, although the method of approximation is quite different in each approach.

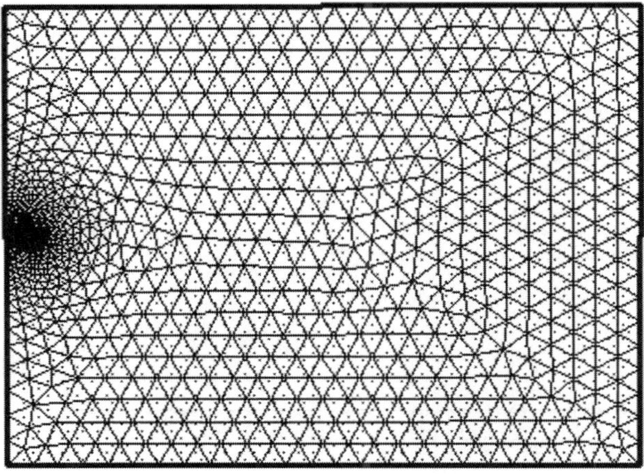

FIGURE 3.69. *Finite-element transport mesh refined in the neighborhood of the source.*

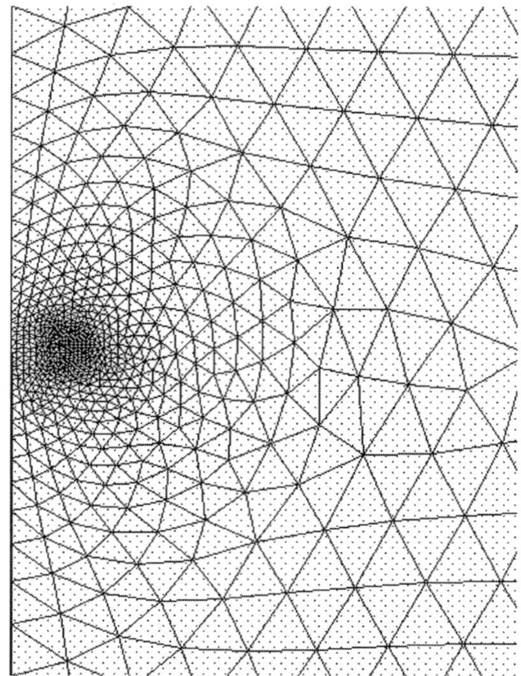

FIGURE 3.70. *Finite element mesh shown in Fig. 3.69 magnified in the area near the source.*

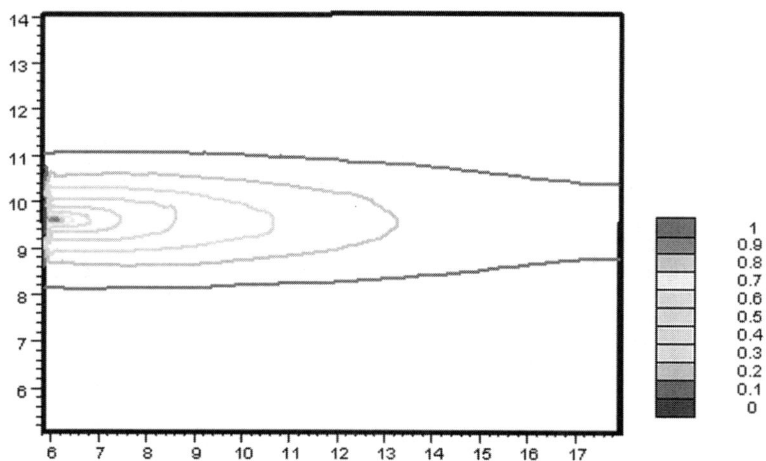

FIGURE 3.71. *Concentration solution obtained using a refined finite-element mesh in the neighborhood of the source. This figure should be compared with Fig. 3.68.*

2. The solutions obtained for flow are very similar, both for trivial one-dimensional flow and for more complex flow, as in the Tucson example.
3. The solution time is approximately the same for both methods, although the number of nodes used in the finite-element model tend to be significantly less than the number used in the finite-difference model.
4. The data input structure is essentially the same in each method, although the *MODFLOW* GUI is, generally speaking, more user-friendly than its *PTC* counterpart.
5. Graphical output is similar in both model GUIs, although the *MODFLOW* and *MT3D* GUIs are a little easier to use.
6. Extension of the model to three space dimensions via modified layer drop-down menus is similar in each model GUI, although the layer-numbering convention is the opposite.

The models differ in the following characteristics:

1. Mesh refinement for the accommodation of singular points such as wells and concentration sources is achieved automatically by the finite-element model GUI.
2. Where mesh refinement is specified by the user, the transition between refined and unrefined areas is gradational using finite-element methods and abrupt using finite-difference models.
3. Irregular boundaries are more easily handled using finite-element methods because of the ability to use triangular elements.
4. The finite-difference GUI allows for the use of a number of third-party packages that facilitate the accommodation of some commonly encountered hydrological conditions.
5. The specification of boundary conditions is more mathematically abstract in the finite-element GUI than in the *MODFLOW/MT3D* GUI, which tends to be more descriptive in terms of a hydrological interpretation.

A comparison of a refined and an unrefined mesh in a transport model revealed that a refined mesh in the neighborhood of the contaminant source resulted in a substantial reduction in numerical diffusion. A comparison of finite-difference, finite-element and MOC3D (a method of characteristics approach) indicated that in a sample problem, both the finite-difference and finite-element models exhibited some numerical diffusion.

Index